Petr N. Vabishchevich (Ed.)
Computational Technologies
De Gruyter Graduate

Computational Technologies

A First Course

Edited by
Petr N. Vabishchevich

DE GRUYTER

Mathematics Subject Classification 2010
35-01, 65-01, 65M06, 65M22, 65M50, 65N06, 65N22, 65N50, 68-01, 68N15, 68N20

Editor
Prof. Dr. Petr N. Vabishchevich
Nuclear Safety Institute
Russian Academy of Sciences
B. Tulskaya 52
Moscow
115191
Russia
vabishchevich@gmail.com

ISBN 978-3-11-035992-3
e-ISBN (PDF) 978-3-11-035995-4
e-ISBN (EPUB) 978-3-11-039103-9

Library of Congress Cataloging-in-Publication Data
A CIP catalog record for this book has been applied for at the Library of Congress.

Bibliographic information published by the Deutsche Nationalbibliothek
The Deutsche Nationalbibliothek lists this publication in the Deutsche Nationalbibliografie;
detailed bibliographic data are available on the Internet at http://dnb.dnb.de.

© 2015 Walter de Gruyter GmbH, Berlin/Munich/Boston
Typesetting: PTP-Berlin, Protago TEX-Production GmbH, www.ptp-berlin.de
Printing and binding: CPI books GmbH, Leck
♾Printed on acid-free paper
Printed in Germany

www.degruyter.com

List of contributors

Dr. Nadezhda M. Afanasyeva
Centre of Computational Technologies
North-Eastern Federal University
Yakutsk
Russia
E-mail: afanasieva.nm@gmail.com

Victor S. Borisov
Centre of Computational Technologies
North-Eastern Federal University
Yakutsk
Russia
E-mail: refactorman@gmail.com

Dr. Aleksandr G. Churbanov
Nuclear Safety Institute
Russian Academy of Sciences
Moscow
Russia
E-mail: alexchurban@mail.ru

Dr. Aleksander V. Grigoriev
Centre of Computational Technologies
North-Eastern Federal University
Yakutsk
Russia
E-mail: re5itsme@gmail.com

Alexandr E. Kolesov
Centre of Computational Technologies
North-Eastern Federal University
Yakutsk
Russia
E-mail: kolesov.svfu@gmail.com

Petr A. Popov
Centre of Computational Technologies
North-Eastern Federal University
Yakutsk
Russia
E-mail: pa.popov@s-vfu.ru

Ivan K. Sirditov
Centre of Computational Technologies
North-Eastern Federal University
Yakutsk
Russia
E-mail: sirditov@gmail.com

Prof. Dr. Petr N. Vabishchevich
Nuclear Safety Institute
Russian Academy of Sciences
Moscow
Russia
E-mail: vabishchevich@gmail.com

Dr. Maria V. Vasilieva
Centre of Computational Technologies
North-Eastern Federal University
Yakutsk
Russia
E-mail: vasilyevadotmdotv@gmail.com

Dr. Petr E. Zakharov
Centre of Computational Technologies
North-Eastern Federal University
Yakutsk
Russia
E-mail: zapetch@gmail.com

Contents

Preface —— ix

Introduction —— xi

Victor S. Borisov and Petr N. Vabischchevich
1   An introduction to C —— 1
1.1   Syntax —— 1
1.2   Constants and variables —— 3
1.3   Expressions and operators —— 6
1.4   Statements —— 9
1.5   Arrays —— 13
1.6   Pointers —— 15
1.7   Functions —— 17
1.8   Structures —— 20
1.9   Input/output —— 21
1.10   Program structure —— 25

Victor S. Borisov and Petr N. Vabishchevich
2   C++ programming basics —— 29
2.1   Enhanced features —— 29
2.2   Classes —— 32
2.3   Function and operator overloading —— 36
2.4   Inheritance —— 38
2.5   Input/output —— 41
2.6   Standard libraries —— 47

Victor S. Borisov and Petr N. Vabishchevich
3   GCC distilled —— 51
3.1   General information —— 51
3.2   Available documentation —— 53
3.3   Compilation workflow —— 60
3.4   Compiling a C/C++ program —— 65
3.5   Libraries and linking programs —— 67
3.6   Debugging —— 70
3.7   The make utility —— 74

Petr N. Vabishchevich
4   The Eclipse IDE in a nutshell —— 79
4.1   The Eclipse architecture and GUI —— 79

4.2   Working with projects —— **86**
4.3   Editing a source code —— **93**
4.4   Debugging applications —— **99**

Nadezhda M. Afanasyeva, Victor S. Borisov, Maria V. Vasilieva, Aleksandr V. Grigoriev, Petr E. Zakharov, Petr A. Popov, and Ivan K. Sirditov

5   The GSL scientific library —— 103
5.1   Preliminaries —— **103**
5.2   Functions and constants —— **104**
5.3   Combinatorics —— **110**
5.4   Linear algebra —— **118**
5.5   Numerical differentiation and integration —— **129**
5.6   Cauchy problem for systems of ODEs —— **139**
5.7   Solution of equations —— **148**
5.8   Interpolation and approximation of functions —— **164**
5.9   Fast Fourier transforms and Chebyshev approximations —— **174**

Ivan K. Sirditov

6   Visualization of computed data —— 181
6.1   Basics —— **181**
6.2   Data handling —— **188**
6.3   Data interpolation —— **191**
6.4   Using styles —— **195**
6.5   Graph decoration —— **202**

Aleksandr G. Churbanov, Alexandr E. Kolesov, and Petr N. Vabishchevich

7   Mathematical modeling —— 209
7.1   Approximation of experimental data —— **209**
7.2   Predator-prey system —— **213**
7.3   Water infiltration —— **219**
7.4   Torsion of cylindrical bars —— **227**

Bibliography —— 233

Index —— 235

Preface

Modern scientific and engineering computations are based on a numerical study of applied mathematical models. Mathematical models involve linear or nonlinear equations as well as systems of ordinary differential equations (ODEs). But in the majority of cases, mathematical models consist of systems of partial differential equations (PDEs), which may be time-dependent as well as nonlinear and, moreover, they may be strongly coupled with each other. In addition, these equations are supplemented with the appropriate boundary and initial conditions. To obtain high-fidelity numerical results of practical interest, it is necessary to find solutions of boundary value problems in complicated computational domains.

The existing literature that discusses problems of scientific and engineering computations, in fact, does not reflect the up-to-date realities. Practically, we have books on numerical methods somehow adapted to the needs of the reader. They focus on scientific and engineering computations from the viewpoint of specialists in numerical methods and computational mathematics. This approach includes developing a numerical method, programming a code, performing computations, and processing numerical results; all of these steps are carried out by the readers themselves. This methodology is appropriate for solving rather simple problems and suggests the sufficient universality of the reader, which is quite rare.

Applied software must reflect the state-of-the-art in numerical methods, programming techniques, and the efficient use of computing systems. This can be achieved with a component-based software development. In this approach, after a modular analysis, a mathematical model is divided into basic computational subproblems, and then an algorithmic interface is organized between them. The solution of the subproblems is implemented using standard computational components of scientific and engineering software. Software packages and modules for pre- and post-processing of problem data may also be treated as components of the developed applied software. The above-mentioned problem-oriented computational components are oriented towards solving typical problems of computational mathematics, and they are developed by experts in numerical methods and programming techniques. The last condition ensures the quality of the developed product when we employ modern computing systems.

Applied software is developed using certain standards and agreements. It concerns, in particular, a programming language. For a long time the software for scientific and engineering computations was implemented in the programming language Fortran. The main advantage of Fortran is a large number of programs and libraries written in it, which are often freely available with a source code and documentation. Nowadays, the situation is changing in favour of other programming languages, especially in favour of C and C++. At present, new mathematical libraries and particular components are usually written in C/C++. Moreover, many well-proven applied software projects developed early in Fortran were rewritten in C/C++.

In research projects, we traditionally focus on using free and open source software (FOSS). It is especially suitable for educational process. From our point of view, the natural business model should be based on payment to an educational institution for training potential users of proprietary software, but in reality, it seems to be quite the contrary. The second requirement, in our mind, is portability, i.e. software should work on various hardware platforms and/or operating systems. In more exact terms, programming languages, available libraries, and applied software should be cross-platform.

Another important issue is related to multiprocessor computers. Applied software for multiprocessor computing systems with shared memory (multicore computers) is developed using OpenMP. For systems with distributed memory (clusters), the standard programming technique is MPI. Applied problems that are governed by PDEs can be solved on parallel computers using the library PETSc.

These key thoughts have determined the structure of this book and its general direction of designing modern applied software. We describe the basic elements of present computational technologies that use the algorithmic languages C/C++. The emphasis is on GNU compilers and libraries as well as FOSS for the solution of computational mathematics problems and visualization of the obtained data. This set of development tools in other circumstances might be slightly different, but this does not change the general orientation.

All the questions of the numerical solution of applied problems on parallel computing systems are discussed in the second volume.

The book was prepared by a team of researchers from the Center of Computational Technologies, M. K. Ammosov North-Eastern Federal University, Yakutsk, Russia, and scientists from the Nuclear Safety Institute, Russian Academy of Sciences, Moscow, Russia. We hope that this book will be useful for students and specialists who solve their engineering and scientific problems using numerical methods. We would be grateful for any any constructive comments on the book.

Petr N. Vabishchevich
Moscow and Yakutsk, March 2014.

Introduction

The C language is a procedural general-purpose programming language with modern data structures management mechanisms and a rich set of operators. The language is not oriented to a particular hardware or operating system. It allows to write easily programs, which can be transferred to other platforms without any changes. At the beginning, a brief introduction to C is given with emphasis on its easy use in scientific and engineering computations. The essentials of the language, such as variables, data types, executable statements, functions, arrays, pointers, dynamic memory, and file management, are described.

Then we present some observations on the C++ programming language, which is derived from C. Most of C elements are not modified in C++, and therefore programs in C are usually translated by C++ compilers without any problem. The principal moment is that C is based on the concept of structured programming, whereas C++ is a fully object-oriented language. The C++ language provides flexible and efficient tools for defining custom types (classes). Classes provide encapsulation, dynamic assignment of types, memory management possibilities, and operator overloading.

The GNU Compiler Collection (GCC) compiler suite is used for translating a program from C/C++ to machine code. GCC is an open-source software distributed by Free Software Foundation (FSF). These compilers are standard not only for open source Unix-like operating systems, but also for some proprietary operating systems. Below, for these compilers, we discuss the issues of program compiling, linking, and debugging.

As a rule, software development is conducted using an integrated development environment (IDE). One of the most popular open-source IDEs is Eclipse. A quick guide to Eclipse is presented in this volume. The main features for editing, compiling, debugging, and application assembling are considered.

The GNU Scientific Library (GSL) is widely employed for scientific and engineering computations. The library is written in C. Using GSL, we can solve the standard problems of computational mathematics: operations with vectors and matrices, linear algebra problems, solution of nonlinear equations, numerical differentiation and integration, interpolation, initial value problems for ODEs, and so on. That is why the use of GSL in applications is also presented here.

In the numerical solution of applied problems, special attention is given to performing data analysis and graphics. In particular, special tools are employed for visualization of calculated data. To construct plots for one- and two-dimensional functions, gnuplot is commonly used. We also employ gnuplot in our computational technologies. This software is based on its own system of commands. In addition, interactive work in command lines and implementation of scripts from files are supported. The resulting plots can be displayed on a screen or saved in files written in widely used formats (PNG and EPS). All these issues are discussed in a special chapter.

Finally, the basic features of computational technologies under discussion are illustrated with model problems. Their numerical solution begins with the formulation of the problem. Then an appropriate computational algorithm is selected. All programs are implemented in C/C++, using the GSL library. Gnuplot is employed to visualize the results of computations.

Victor S. Borisov and Petr N. Vabischchevich

1 An introduction to C

Abstract: In this chapter we present a brief description of the basics for the C programming language, which is very popular in scientific and engineering computations. It is assumed that the reader is familiar with the programming fundamentals.

1.1 Syntax

The basic syntax constructions of a C program are described below in order to understand the codes presented in other chapters of the book.

1.1.1 Characters

The following characters are used in C:
- The upper and lower case letters:

```
A B C D E F G H I J K L M N O P Q R S T U V W X Y Z
a b c d e f g h i j k l m n o p q r s t u v w x y z
```

- Decimal numerals:

```
0 1 2 3 4 5 6 7 8 9
```

- Punctuation marks:

```
! " # % & ' ( ) * + , - . / : ; < = > ? [ \ ] ^ _ { | } ~
```

- Whitespace: space, tab, newline and form feed.

Special character combinations called escape sequences are used in C to represent difficult or impossible-to-type characters (see Table 1.1). The first required character of an escape sequence is \, whereas other characters are Latin letters or numbers.

In addition, any character can be presented by a combination of three octal digits using \ddd or \xddd in hexadecimal representation, respectively.

Table 1.1. Escape sequences.

Sequence	Description
\a	Alert
\b	Backspace
\f	Form feed
\n	Newline (line feed)
\r	Carriage return
\t	Horizontal tab
\v	Vertical tab
\'	Single quote
\"	Double quote
\?	Question mark
\\	Backslash
\0	Null character (empty)

In standard C, non-Latin alphabets are not supported. For specific compilers, such an extension is possible. In particular, it is appropriate to employ non-Latin letters in comments.

1.1.2 Comments

Comments help to understand a program code and are ignored by any compiler. Comments can be used anywhere where a blank or newline are allowed. Block comments are delimited by /* and */. Comments are not nested. A single-line comment (introduced in the standard C99) can begin with the character // and end at the end of the line.

Listing 1.1.

```
1  /*
2    Multiline
3    comments
4  */
5  //  One-line comments
6  double pi = 3.145926536;     // the Pi number
```

1.1.3 Identifiers

Identifiers are employed as names of variables, functions, and data types. They are created by a declaration of a variable, function, structure etc.

Upper and lower case letters, digits, and the underscore character _ can be used in identifiers. The first character cannot be a digit. There is no restriction on the length of identifiers, but only the first 31 characters are significant for most compilers.

Keywords are specific reserved identifiers or system words that are used as operators in C. Keywords cannot be utilized as names. The list of the C keywords is given below:

auto	break	case	char	const	continue
default	do	double	else	enum	extern
float	for	goto	if	int	long
register	return	short	signed	sizeof	static
struct	switch	typedef	union	unsigned	void
volatile	while				

The use of an underscore as the first character of an identifier requires some caution. Such identifiers may conflict with the names of system functions and/or variables, and programs with them may have problems with portability.

1.2 Constants and variables

Here we introduce the basic data types that C uses. The syntax of constants and the amount of memory occupied by various types of constants are described below.

1.2.1 Data types

Programs work with data stored in computer memory. The storage and processing of various data types, such as integer and floating-point numbers, are implemented in various ways.

Data can have a prescribed value before running a code and remain unchanged throughout execution. This data is referred to as constants and presented by literals in a code. A variable has a name, and its value can be changed. Thus, a variable is a named memory location, where we can initialize some value and modify it during performing calculations.

A compiler recognizes a constant by a form employed to declare it in a program. To use a variable, we have to declare a name and type of memory allocation in this declaration. The basic **data types** in C are integers (keywords are int, short, long, unsigned), symbols (char) and floating-point numbers (float, double). Any variable can be declared as unchangeable by adding the const keyword. We cannot assign a new value to variables of this type.

1.2.2 Integer constants

The types short, int, and long are designed to represent integers. To declare integers, we must write only a type that is followed by a list of variable names.

Listing 1.2.

```
1  int a,b;      // a,b - integers
```

A comma is used to delimit identifiers.

Decimal integers are represented using the digits 0–9, but 0 cannot be the first digit. Octal integers start with 0 and are represented by digits 0–7. Hexadecimal integers start with 0x or 0X and use digits 0–9 and letters a–f or A–F for representing 10–15.

```
154 = 0252 = 0x9A
```

If a prescribed value is larger than it must be by default, the long type can be used. A long integer constant is explicitly defined by the letter l or L after the constant. The short type is employed to save memory. Using unsigned int, unsigned long, unsigned short, we can handle only positive integers. Octal and hexadecimal constants can also have the modifier unsigned. For this, the letters u or U are used after constants.

In the C language, a range of values for representing integers is not specified. Usually a short integer occupies half of a machine word, int employs one word for storage, whereas long corresponds to one or two words, respectively. For 32-bit machines, the word size is equal to 4 bytes (1 byte is equal to 8 bits), whereas int and long usually have the same size (see Table 1.2).

Table 1.2. Integer data types.

Type	Size (bytes)	Range of value
int	2	−32,768–32,767
	4	−2,147,483,648–2,147,483,647
unsigned int	2	0–65,535
	4	0–4,294,967,295
short	2	−32,768–32,767
unsigned short	2	0–65,535
long	4	−2,147,483,648–2,147,483,647
unsigned long	4	0–4,294,967,295

Table 1.3. Floating-point numbers.

Type	Size	Range of values	Precision
float	4	1.175494351e-38–3.402823466e+38	7
double	8	2.22507385850720e-308–1.79769313486231e+308	15
long double	10	3.36210314311209e-4932–1.18973149535723e+4932	19

The operator `sizeof` is used to determine the size of a type or variable in bytes.

Listing 1.3.

```
1  a = sizeof(int);
2  b = sizeof(long);
3  c = sizeof(short);
4  d = sizeof(unsigned long);
```

1.2.3 Floating-point numbers

The fundamental data type in C is `float`. For computing with double precision, the `double` (`long float`) type is used to represent a double value of bits for number presentation. If `double` is not enough, then we can apply `long double`.

Real numbers are represented by an integer part, a decimal point, fractional part, letter `e` or `E`, and an integer exponent part with an optional sign. Integer and fractional parts are represented as a sequence of digits. Either an integer or a fractional part of a real number may be absent. This is the same for a decimal point or `e` (`E`).

```
-5.2e+7    2.8e-2 3.14    .2    7e12    .8E-5    90.
```

Floating-point numbers can be stored with only limited precision, which is defined by the binary format for the presentation of real numbers. The precision (see Table 1.3) is expressed by the number of significant digits (symbols). In this case, the position of a decimal point does not matter.

Complex numbers are supported starting with the standard C99.

1.2.4 Using characters

A character variable is identified by the keyword `char`. The `char` type defines integer numbers without sign in the range from 0 to 255 (the size of 1 byte), and a character constant consists of one `ASCII` code between single quotes.

Listing 1.4.

```
1  char a = 'A', b = '\065', c = '\x41';   // a = b = c
```

1.3 Expressions and operators

Operators define some operations on data (operands) to be performed by a computer. The essentials of C operators is given below.

1.3.1 Arithmetic operators

Table 1.4 shows the fundamental **arithmetic operators.** A value-assigning operator is the simplest operator.

Table 1.4. Arithmetic operators.

Operator	Operation
=	Assignment
+	Addition
−	Subtraction
*	Multiplication
/	Division
%	Modulo (integer remainder)
+=	Addition with assignment
−=	Subtraction with assignment
*=	Multiplication with assignment
/=	Division with assignment
++	Increment
−	Decrement

Listing 1.5.

```
1  float a, b = 5.;
2  a = b;        // a = 5.
3  b = b + 1.;  // b = 6.
```

The assignment operators presented below are often employed to update variables; they reduce a code and make it more convenient in use. For instance, the statement `n = n + 5` can be written as `n += 5`. Such a combination is also possible for other arithmetic operators.

Listing 1.6.

```
1  float a = 8.;
2  a += 1.;     // a = 9.
3  a -= 3.;     // a = 6.
4  a *= 2.;     // a = 12.
5  a /= 4.;     // a = 3.
```

In C, any fractional part resulting from integer division is discarded. The expression `n % m` yields an integer remainder of dividing n by m.

Listing 1.7.

```
1  int l, n = 5, m = 3;
2  l = n / m;   // l = 1
3  l = n % m;   // l = 2
```

The `++` operator increases the value of its operand by 1, whereas `--` decreases its operand by 1. Thus, instead of `n = n + 1`, we can write `++n`. The increment and decrement can precede an operand (prefix form) or follow it (postfix form). If the increment (or decrement) precedes an operand, this operation is executed before the result is used in the expression. If the increment operator follows an operand, the operation is executed after returning the value of the operand.

1.3.2 Relational and logical operators

Relational operators are applied when the values of two variables are compared with each other. Logical operators provide formal logic operations. Results of relational operators are often stand as operands of a **logical operator**.

In the C language, various operators are employed to compare operands of any type (see Table 1.5). Logical operators are listed in Table 1.6.

The values `true` and `false` are used as operands and results of relational and logical operators. The `true` value is represented by any number other than 0, whereas `false` is equal to 0. The result of relational or logical operators is `true` (1) or `false` (0).

The operators `=>`, `>`, `=<`, `<` have the same precedence; the equality/inequality operators `==`/`!=` follow immediately behind them in the precedence order. The precedence of the operator `&&` is higher than of `||`, and both are lower than relational and

Table 1.5. Relational operators.

Operator	Operation
>	Greater than
>=	Greater than or equal to
<	Less than
<=	Less than or equal to
==	Equal to
!=	Unequal to

Table 1.6. Logical operators.

Operator	Operation
&&	AND
||	OR
!	NOT

equality operators. Round brackets are used to change the order of executing relational and logical operations.

Listing 1.8.

```
1  float x = 10.;
2  int l, m,  n = 2;
3  l = (x > 6 || x < 4) && n == 1; // l = 0 (false)
4  m = x > 6 || x < 4 && n == 1;        // m = 1 (true)
```

The arithmetic operators + and – have the same precedence, which is lower then the precedence level of the operators $*$, /, and %, which is in turn lower than logical operators.

1.3.3 Type conversions

Statements and expressions should normally employ variables and constants of just one type. If variables of different types appear in an operator, **type conversions** occur. In the assignment operator, the value of the variable at its right is converted to the type of the variable at its left if it is possible. Each char variable is converted to int, int to float, and float to double. In this case, precision of computations does not increase, but just a representation of numbers is changed. For example, the inverse conversion from double to float loses precision.

Listing 1.9.

```
1  char ch;
2  int n = 75;
3  float a = 123.456;
4  ch = n; // ch = 'K'
5  n = a;  // n = 123
6  a = ch; // a = 75.
```

One value can be assigned to many variables in an assignment operator.

Listing 1.10.

```
1  float a, b;
2  int n = 1;
3  a = b = n + 1;  // a = b = 2.
```

Also, the explicit conversion of variable types is possible.

Listing 1.11.

```
1  int x = 1;
2  float y = sin((float) x);
```

In the above example, the function sin takes only arguments with the float or double type. If an argument has the int type, it is necessary to convert it explicitly to the float type.

1.4 Statements

A statement is considered as a part of a program, which can be executed separately. In C, we have control-flow statements that specify the order in which computations are performed.

1.4.1 Conditional statements

In the C language, if and switch are conditional statements for making a decision.

There are two forms of `if`:

```
if (expression)
    statement;
```

and

```
if (expression)
    statement;
else
    statement;
```

In this construction, a statement may be one operator, a block of operators or empty. A compound statement (block) consists of one or more operators enclosed in braces { }. An empty statement has only a semicolon ; .

An expression is evaluated and if it is true (non-zero), then the statement following `if` is executed. Otherwise the statement following `else` is executed (see the second form of the `if` statement).

A statement `if` can be inside other `if` or `else` statements. An example is the following nested `if` sequence called a `if-else-if ladder`.

```
if (expression)
    statement;
else if (expression)
    statement;
else if (expression)
    statement;
. . .
else
    statement;
```

In this case, the expressions are evaluated from the top down. If some expression is satisfied, the statement associated with this expression is executed, and the rest of a `ladder` is skipped.

Instead of `if-else`, a ternary operator `?:` is often used. The following constructions are equivalent:

```
if (condition)
    variable = expression1;
else
    variable = expression2;

variable = condition ? expression1 : expression2;
```

The switch statement is employed to select a branch of a computational process based on a value of a control expression. The control expression of switch should be expressed as an integer (int or char).

```
switch (expression)
{
    case constant1:
        statements
        break;
    case constant2:
        statements
        break;
    default:
        statements
        break;
}
```

The value of a control expression is compared with constants in case operators. If the value of the switch expression matches one of the constants, the control is transferred to the corresponding mark case and the statements before break are executed. An operator break provides an immediate exit from switch.

1.4.2 Loop statements

Loop statements are intended for repeated execution of a sequence of operations until some condition is satisfied. The condition can be preset before a loop (the for loop) or varies during the execution of a loop body (the while and do-while loops).

The basic form of the for loop is shown here.

```
for (initialization; condition; increment)
    statement;
```

In initialization, an initial value of a variable (loop parameter) is assigned. The condition determines whether or not to execute a statement (body of this loop) again. An update of a loop parameter is performed at each iteration in increment. A for loop is executed if the condition is satisfied, otherwise the loop is terminated.

Listing 1.12.

```
1  // factorial calculation
2  long nFact = 1;
3  int i, n = 10;
4  for (i = 0; i < n; i++)
5       nFact *= i + 1;
```

In standard C99, it is possible to declare a variable in the initialization section of a for loop. This variable is local, and its scope extends to the body of the loop.

Listing 1.13.

```
1  // calculation of e
2  double x = 1.0, e = 1.0;
3  int n = 10;
4  for (int i = 0; i < n; i++) {
5       x = x * (i + 1);
6       e = e + 1 / x;
7  }
```

Any section of a for loop can be omitted and an infinite loop can be obtained if all sections are empty.

```
for( ; ; )
    statement;
```

In the while loop which has the form

```
while (condition)
    statement;
```

any valid expression can serve as a control expression. If a condition is true (the value of the expression is nonzero), then the statement of this loop is executed. Otherwise the loop is terminated.

In the for and while loops, a condition is checked before an iteration. In the do-while loop, such testing is performed after performing an iteration.

```
do {
    statement;
} while (condition);
```

This loop is executed until the condition is true. Braces are not required if a statement is not a block.

1.4.3 Jump statements

In C, the following jumps are determined: break, continue, return, and goto. The statements break and continue can be used in any loops, and moreover, break can be employed in switch. A return statement is applied anywhere within a function, and goto is applied anywhere in a program.

Applications of break in switch are discussed above during consideration of conditional statements. In loops, a break statement leads to unconditional termination of loop execution with the transition to a statement after this loop.

Listing 1.14.

```
1  int i, n = 0;
2  for (i = 1;; i++) {
3        n += i * i;
4        if (n >= 1000)
5              break;
6  }
```

A break statement interrupts a loop, whereas continue only provides the interruption of the current iteration, and the jump to the next one. Control is transferred to the beginning of the nearest outer statement of a while, do or for loop.

A goto jump statement transfers control to a statement with a label.

```
goto label;
 . . .
label: statement;
```

A label is an identifier followed by a colon. It must be anywhere within the function, where goto is used (after or before the goto statement).

1.5 Arrays

Arrays are the fundamental objects used in scientific and engineering computations. An array is a series of values of the same type stored sequentially. The whole array bears a single name. The access to an array element is performed using an integer index.

1.5.1 One-dimensional arrays

The index of the first element of an array is equal to zero. Any array is located in a separate continuous memory domain. The size of any array is defined by a constant. The first element of an array is located in the memory with the lowest address, and the last element corresponds to the biggest address.

The initialization of an array is conducted using a list of values, which is a list of constants separated by commas.

Listing 1.15.

```
1  // Calculation of the maximum and minimum values
2  int n = 7;
3  int i;
4  float a[7] = {7.3, 8.5, 1.3, 5.4, 6.2, 3.4, 5.9};
5  float sMin, sMax;
6  sMin = sMax = a[0];
7  for (i = 1; i < n; ++i) {
8      if (a[i] < sMin)
9          sMin = a[i];
10      if (a[i] > sMax)
11          sMax = a[i];
12  }
```

1.5.2 Character arrays

An example of a one-dimensional array is a string that is a character array ending with the special null character ('\0'). A string constant is a sequence of ASCII characters between quotes.

To declare a character array intended for storing a string, it is necessary to provide a place for the null character, i.e. the size of the array should be 1 more than the number of characters in the string.

If a size of an array is not defined during initialization, then the size is determined automatically by the number of characters.

Listing 1.16.

```
1  // Character array initialization
2  char str1[5] = { 't', 'e', 's', 't', '\0' };
3  char str2[5] = "test";
4  char str3[] = "test";
```

1.5.3 Two- and multidimensional arrays

For multidimensional arrays, as many pairs of brackets as the dimension of the array are used. Numbers in brackets are sizes of an array in the selected dimension. A **two-dimensional array** a[n][m] can be treated as a matrix with n rows and m columns, where n and m are constants.

When handling multidimensional arrays, a computer spends a lot of time on the calculation of addresses. Because of this, access to elements of a multidimensional array is much slower than access to elements of a one-dimensional array.

Multidimensional arrays are initialized in the same way as one-dimensional ones. In a multidimensional array, the size of the leftmost dimension can be omitted. The size of other dimensions must be explicitly defined.

Listing 1.17.

```
1   // squares of integers
2   int nSqrs[][2] = {
3       1, 1,
4       2, 4,
5       3, 9,
6       4, 16,
7       5, 25,
8       6, 36,
9       7, 49,
10      8, 64,
11      9, 81,
12      10, 100
13  };
```

1.6 Pointers

A **pointer** is a variable whose value is the address of an object in computer memory.

1.6.1 Pointer declaration

A pointer declaration consists of a type, the asterisk character * and a variable name.

```
type *name;
```

A pointer of any type can refer to any place in memory, but operations performed with pointers significantly depend on its type. The special character NULL indicates an empty pointer.

Listing 1.18.

```
1  int *pN;
2  char *pCh;
3  float *pFl = NULL;
```

1.6.2 Pointer operations

There are two basic pointer operations, * and &. The & operation applied to some variable returns the address of this variable, but not its current value. The * operation returns the value of the variable located at the specified address.

Listing 1.19.

```
1  int nNumber;
2  int *pPointer;
3  // variable initialization
4  nNumber = 7;
5  pPointer = &nNumber;
6  // change nNumber by pPointer
7  *pPointer = 13; // nNumber = 13
```

1.6.3 Pointers and arrays

In C, arrays and pointers are closely interrelated. Operations with indexes of arrays can be replaced by operations with pointers.

Listing 1.20.

```
1  int A[5] = {1, 2, 3, 4, 5};
2  int x, y, z;
3  int *pA;
4  pA = A;
5  x = *pA;           // x = A[0]
6  y = *(pA + 2);     // y = A[2]
7  z = *pA + 5;       // z = A[0] + 5
```

We can consider the name of an array as the pointer that refers to the first element of the array (x = *pA).

The calculation of y in the above example illustrates arithmetic operations with pointers, i.e. the pointer to the first element is added to 2. The result is the address of the first element plus the memory size of two elements of the array, i.e. the pointer to A[2].

1.7 Functions

A **function** is an independent unit of a program created to resolve specific tasks. In C, functions play the same role as subroutines or procedures in other algorithmic languages. They are the minimum executable entities of a program. Functions are used in order to break large tasks into smaller subtasks which can be executed many times during runtime. A function must be declared before its use.

1.7.1 Function definition

The basic form for defining a function returning a value is

```
type function_name(parameters list) {
    body of function
}
```

In the function header, type defines the type of a returning value (int, double, char etc.) or void if the function returns nothing. If a type is not indicated, it is assumed that the function returns int.

A function name should be unique and must not be identical with the keywords or other functions of a program.

A list of parameters is either empty or it includes arguments separated by commas and has the form

```
type parameter_1, type parameter_2, ..., type parameter_n
```

The body of the function is restricted by braces and includes compound statements or blocks. A block differs from a compound statement in that it can include definitions of other objects, e.g. variables or arrays. In C, a function cannot be defined within the body of another function (a definition of a function cannot be nested).

In a body of a function statement, `return` is used to provide immediate exit from the function and jump to a caller. Two forms are used, i.e.

```
return;
return expression;
```

The first form is employed for functions whose returned type is `void`. In the second form, the expression has the type declared in the function definition.

It is possible to omit the `return` statement. A compiler automatically adds it to the end of the function's body.

An example of a function definition is as follows:

Listing 1.21.

```
1  // function definition
2  float f(float a) {
3      if (a > 0)
4          return a;
5      else
6          return -a;
7  }
```

1.7.2 Function prototype

To call a function before it is defined in the current file or if it is described in a different file, it is necessary to put a function declaration (prototype). This gives compilers an opportunity to check the types of arguments passed to the function with the types of parameters in the function definition.

A function prototype has the same form as a function definition but without a function body; it ends with a semicolon.

```
type function_name(parameters list);
```

A simplified form of a function prototype does not include the names of parameters; only the parameter types are specified, whereby the prototypes of the same function are equivalent.

Listing 1.22.

```
1  void f(int n, float a);
2  void f(int, float);
```

An example of a function call:

Listing 1.23.

```
1  // function prototype
2  float f(float a);
3  . . .
4  float a, b;
5  a = sin(10.);
6  // function call
7  b = f(a);
```

1.7.3 Pointers as function formal parameters

In C, function parameters are passed by value, and thus a called function cannot change a parameter value by executing statements in the body of the function. This restriction can be overruled if the addresses of variables are passed to a function as actual arguments.

An important case is when a name of an array is used as a function formal parameter. In this case, only the address of the beginning of the array is passed to a function, whereas the array elements are not copied. Thus, in a body of a function, we can change the elements of arrays. There are three forms of passing arrays to functions:
- the parameter is defined as an array with an indication of dimension;
- the parameter is defined as array without indication of dimension;
- the parameter is defined as a pointer.

The following is an example of equivalent specifications of an array:

Listing 1.24.

```
1  void f(float a[100]);
2  void f(float a[]);
3  void f(float *a);
```

1.8 Structures

A **structure** is a collection (set) of one or more variables, possibly of different types, which are grouped together under the same name. Structures allow grouping of connected data and their operation as a unit rather than as separate entities.

1.8.1 Structure definition

When creating a structure, we introduce our own data type. A structure must have a name, and each member of a structure must also have a name and a type.

```
struct structure_name {
    type_1 member_name_1;
    type_2 member_name_2;
    . . .
    type_n member_name_n;
};
```

In the following example, the structure `point` defines the position of a point on a plane.

Listing 1.25.

```
1  // Structure definition
2  struct point {
3      double x; // x coordinate
4      double y; // y coordinate
5  };
```

1.8.2 Structure variables and their initialization

A structure definition creates a new data type, which can be used to declare variables. Structure variables are declared in the same way as variables of other types.

Similarly to arrays, structures can be initialized by using an initialization list. The following is an example of the definition and initialization of the `point` structure variables:

Listing 1.26.

```
1  // initialization
2  struct point a = { 0, 1 }, c = { 3, 5 }, *b = &c;
```

1.8.3 Gaining access to structure members

To access a structure member, the member with the variable name must be specified using the following form:

```
struct_name.member_name
```

If a pointer to a structure is used to access structure members, the -> operator is employed. The following is an example of the access to a member of the point structure:

Listing 1.27.

```
1  double d;
2  . . .
3  // interval length
4  d = sqrt((b->x - a.x) * (b->x - a.x) + (b->y - a.y) * (b->y - a.y));
```

1.9 Input/output

Input and output operations in C are implemented by the standard input/output library (the header file is stdio.h) using functions scanf, printf, fscanf, and fprintf. The functions fopen and fclose provide opening and closing files.

1.9.1 Library functions

The C language is accompanied by a set of standard library functions which perform various tasks. In particular, the operations of input and output are realized via library functions. Prototypes of library functions are contained in special header files supplied with libraries as a part of programming systems. To use library input-output functions, we apply the directive

```
#include <stdio.h>
```

1.9.2 Streams

The information input from peripheral devices (such as a keyboard) is stored in random access memory (RAM), and output information is directed from RAM to peripheral devices (the display, printer, etc.). A hard disk drive can be employed both for input and output. The process of information exchange between RAM and the peripherals is provided by streams. A stream is a byte sequence transmitted during an input/output process.

There are standard streams and streams that are connected with files on a disk. Standard streams are created and opened by the system automatically. The main streams are
- `stdin` – the standard input stream (connected with a keyboard);
- `stdout` – the standard output stream (connected with a video display).

In C, streams are of two types: text and binary. A text stream is a sequence of characters. When transmitting characters from a stream to a display, some of them are not shown (e.g., carriage return, newline, etc.). A binary stream is a sequence of bytes which clearly correspond to what is on an external device.

1.9.3 Work with files

To work with a **file**, we begin with a declaration of a pointer to a stream using the form

```
FILE *pointer_name;
```

Opening a stream is executed by the standard function `fopen` defined as

```
FILE *fopen(const char *filename, const char *opentype);
```

If for some reason during stream opening an error occurs, then the `fopen` function returns the value `NULL`.

How to open a file is shown as follows:

Listing 1.28.

```
1  FILE *pF;
2  pF = fopen("test.dat", "w");
```

Here the first parameter `test.dat` is the file name in the current directory associated with the stream with the `pF` pointer. The second parameter `w` means that the file is opened in write mode.

Table 1.7 presents the possible opening modes.

When working with binary files, the `b` letter is added to the opening mode (e.g. `rb`, `wb`, `r+b`, `ab+`).

After finishing with a file, it should be closed. Closing a file is performed by the library function `fclose`, which has the following prototype:

Table 1.7. Opening modes.

Parameter	Mode
r	Open for reading
w	Open for writing (the previous contents disappears)
a	Open for append
r+	Open for both reading and writing
w+	Create for both reading and writing
a+	Open to append or to create for both reading and writing

```
int fclose(FILE *stream)
```

If closing is successful, then `fclose` returns zero; other values indicate an error.

1.9.4 Formatted data output

Using the `fprintf` function of formatted output, we can generate a text file with computational results in character form. The `printf` function is designed to cause output to a display. This function is a particular version of `fprintf`, which works with the standard stream `stdin`.

The description of functions to display and write to the stream a **format string** is

```
int printf(const char *template, ...)
int fprintf(FILE *stream, const char *template, ...)
```

For example, we can print a string as follows:

Listing 1.29.

```
1  printf("string 1\n");
2  fprintf(stdout, "string 2\n");
```

A string can contain formatting specifications, which display a subsequent argument from a list. Specifications begin with the character %. The following is an example of printing numbers with comments.

Listing 1.30.

```
1  int n = 5;
2  double a = 3.1415926535898;
3  printf("n = %d, a = %15.10f, a/4 =  %f", n, a, a/4);
```

The result is

```
n = 5, a 1= 3.1415926536, a/4 = 0.785398
```

The format %d (also, %i) is used for integers written in decimal form. To work with float variables, the %f format is applied, which by default outputs 6 digits after the comma. In the format %15.10f, 15 positions are prescribed to represent a number, and 10 of them are for digits after the decimal point.

The formats %e and %E are employed to output floating numbers in exponential notation. Using the command

Listing 1.31.

```
1 printf("a = %e, a/4 =  %E", a, a/4);
```

we get

```
a = 3.141593e+000, a/4 = 7.853982E-001
```

The %s format is applied to output a string.

1.9.5 Formatted data input

The scanf (fscanf) function which provides input is an analogue of printf (fprintf), but it performs formatting in the opposite direction. The function prototypes are

```
int scanf(const char *template, ...)
int fscanf(FILE *stream, const char *template, ...)
```

The **scanf function** reads characters from the standard input stream, interprets them according to string specification template, and sends results to its other arguments. The scanf function ends the work if the format is concluded or an input value does not conform with the controlling specification. The scanf function returns the number of successfully entered data elements as the result. If a file is concluded, the result is EOF.

An example of finding the sum of two numbers entered from a keyboard:

Listing 1.32.

```
1  int a, b;
2  printf("Enter two numbers:\n");
3  scanf("%d %d", &a, &b);
4  printf("Sum of the two numbers = %d\n", a + b);
```

1.10 Program structure

A C program is a set of functions and data declarations contained in one or more files, which are called source files. One of these functions is the main function. Files are compiled independently from each other and built with procedures from libraries forming an object code of the program. Such separate compilation results in the fact that any modification of one file does not require recompiling of the entire program.

1.10.1 Preprocessing facilities

A preprocessor is included in compilers as a mandatory component. Its purpose is to process the source files of a program before compilation. The preprocessor modifies a source file, and only after that the file is compiled. Preprocessing includes, in particular, the replacement of identifiers by prescribed character sequences, the inclusion of file texts, the exclusion of individual parts of the code, conditional compilation, etc.

Preprocessor directives are written on a new line, and the first character must be #. Preceding gaps and tabs are ignored. The end of a directive is the end of a line; if the directive does not fit on one line, the character \ is placed at the end of the line and the directive continues on the next line. The directives #include and #define are examples of preprocessing facilities.

The #include directive creates a copy of a file mentioned in it, and adds it to our program. There are two forms of using #include. In the first case, we write

```
#include <file_name>
```

and searching for the file is conducted in the system folders where library files are located. If we use the directive

```
#include "file_name"
```

then searching is carried out in the current folder, where the source files are located.

The `#define` directive serves to replace often used constants, keywords, operators, or expressions by some identifiers. The directive syntax is

```
#define identifier value
```

Before compilation, in all places where the identifier is used, it is replaced by the proposed value.

An example is as follows:

Listing 1.33.

```
1  #define PI 3.14159265
```

1.10.2 File organization of a program

In the C language, source files are of two types: headers (with the `.h` extension) and source code files (with the `.c` extension). Header files serve to transfer information between modules and contain only descriptions, i.e. the necessary information is already written in the blocks of a program. Mainly, this is function prototypes, which describe function names, returned variable types, types of arguments. In headers, there are also described names and types of external variables, constants, structures, etc. For instance, the description

```
extern int m;
```

means that the variable m of the integer type is defined somewhere in some source code file of a project. In this case, the description provides information about the external variable, but does not define the variable itself.

Source code files are separate modules developed and compiled independently and combined during creation of an executable program. Such files can include descriptions contained in header files. In turn, header files can also use other header files. The `#include` directive is used to include header files.

1.10.3 Program structure

A C program consists of one or more functions. It is necessary to define only the `main` function, where a program begins execution. The `main` function always contains statements (mainly function calls), which reflects the essence of the problem to be solved.

The structure of a typical C program may be written as follows:

```
Header files inclusions
Global variable declarations
Function prototypes
int main() {
    statements
}
type_1 function_name_1 (parameters list) {
    statements
}
type_2 function_name_2 (parameters list) {
    statements
}
. . .
type_n function_name_n (parameters list) {
}
```

To illustrate the general construction above, let us consider a program for the square root calculation by Heron's iterative method.

Listing 1.34.

```
1  #include <stdio.h>                  // header i/o file
2  const double eps = 0.0000001; // precision
3  double Sqrt(double x); // function prototype
4  //  main function
5  int main() {
6      double x;
7      double y;
8      printf("Enter x: ");
9      scanf("%lf", &x);
10     x = 2.0;
11     y = Sqrt(x);
12     printf("sqrt(x) = %e", y);
13     return 0;
14 }
15 // square root calculation function
16 double Sqrt(double x) {
17     double xn;
18     xn = x;
19     // Heron's iterative method
20     do {
21         xn = 0.5 * (xn + x / xn);
22     } while (xn * xn - x > eps);
23     return xn;
24 }
```

After executing this program, we obtain the following output:

```
Enter x: 2.
sqrt(x) = 1.414214e+000
```

In this program, the %lf format is employed in scanf to input numbers with double precision.

Victor S. Borisov and Petr N. Vabishchevich

2 C++ programming basics

Abstract: New features of the **C++ programming language** in comparison with C are associated with the support of abstract data types and object-oriented programming. A class is the most important concept of C++.

2.1 Enhanced features

Below we consider some new possibilities of C++ (in comparison with C) not connected with the paradigm of object-oriented programming.

2.1.1 Keywords

The list of keywords of C (page 3) is extended in C++. The following keywords are reserved in addition to C:

catch	class	delete	friend	inline	new
namespace	operator	private	protected	public	template
this	throw	try	using	virtual	

2.1.2 Declaration of variables within blocks

In C, a declaration of local variables within a block is placed at the beginning of the block, before the first executable statement. As for C++, variables can be declared anywhere within a block before their use. The declaration and use of a loop counter is shown in the following example.

Listing 2.1.

```
1  // calculation of Pi = 3,141592653589793238...
2  double Pi = 2.0;
3  for (int i = 2; i < 10000000; i += 2) {
4      double s1;
5      s1 = i / (i - 1.0);
6      double s2;
7      s2 = i / (i + 1.0);
8      Pi *= s1 * s2;
9  }
10 printf("Pi = %15.12f\n", Pi);
```

The local variables s1, s2, i are not available outside the loop.

2.1.3 Scope resolution operator

The scope resolution operator : : provides access from a block to a global variable which has the same name as a local variable. For example,

Listing 2.2.

```
1  int i = 1; // global variable
2
3  int main() {
4      int i = 2; // local variable
5      int ig;
6      ig = ::i;
7      printf("global i = %d, local i = %d", ig, i);
8      return 0;
9  }
```

The output is

```
global i = 1, local i = 2
```

2.1.4 Default arguments

In C++, arguments can be omitted in function calls. In this case, default values are used for the omitted arguments. Default values are defined in a function prototype. If a value of an argument is omitted, then the values of all the following parameters must also be omitted.

Listing 2.3.

```
1  // function prototype
2  double f(double x, int n = 2, int m = 10);
3  . . .
4  double y;
5  // function calls
6  y = f(1.0);        // f(1.0, 2, 10) is called
7  y = f(1.0, 3);     // f(1.0, 3, 10) is called
8  y = f(1.0, 3, 12); // f(1.0, 3, 12) is called
```

2.1.5 Boolean type

In C, there is no specific logical type; integer variables are used instead. The true value is associated with any nonzero integer; false corresponds to zero.

C++ directly introduces the boolean type bool. A **boolean variable** takes two values: false and true.

An example of boolean variables declaration:

```
bool a, b = true;
```

2.1.6 Dynamic memory

In C++, an object can be created (declared) in any place where it needs to be used. The operators new and delete are used for this. The following example illustrates a typical situation of dynamic creation of an array.

Listing 2.4.

```
1  int n = 10;
2  int *a = new int[n];
3  for (int i = 0; i < n; i++) {
4      a[i] = i * i;
5      printf("%d %d \n", i, a[i]);
6  }
7  . . .
8  delete[] a;
```

Memory allocation is performed by the statement new int[n], whereas deallocation is carried out via delete[] a.

In C, the direct application of library **functions malloc, calloc**, and **free** is required. For the above-mentioned example, the header file stdlib.h of the C standard library must be attached, and two lines must be changed.

Listing 2.5.

```
1  // memory allocation
2  a = (int *) malloc(n * sizeof (int));
3  . . .
4  // memory deallocation
5  free(a);
```

2.2 Classes

C++ gives the programmer the opportunity to expand independently the capabilities of the programming language by creating his own data type required for solving specific problems. A **class** is a collection of related variables and associated functions intended to work with variables. Classes in C++ are an advanced version of structures (struct) in C.

2.2.1 Creating a class

A class declaration defines a new data type. A class is logical abstraction, and an object is an instantiation (a copy of class). A declaration of a class is similar to a declaration of structure.

```
class class_name {
    closed types and functions
access_specifier_1:
    data and functions
access_specifier_2:
    data and functions
    . . .
access_specifier_n:
    data and functions
} object_list;
```

An object list allows the declaration of the class objects and can be absent.

Each class contains data (data members) and a collection of functions (methods) to handle this data. Declarations can include both methods and data declarations.

A **class definition** includes access specifiers besides declarations of data and functions. Access specifiers specify scope rules (visibility) for members of the class:
- private – members are visible only within their class and cannot be used outside;
- public – members are visible for both members of the class and the outside code;
- protected – members are visible for other members and children of the class (in inheritance).

Encapsulating in object-oriented programming is the hiding of class realization details. Using an access specifier, a class introduces another scope besides a file, function, block, function prototype.

By default, methods and data declared in a class are private and accessible only with methods of this class. An access specifier extends its functioning up to the next specifier or the end of the class definition.

This is the class of complex numbers.

Listing 2.6.

```
1  // Class definition
2  class Complex {
3  private:
4      double real; // real part
5      double image; // image part
6  public:
7      double module(); // module of a complex number
8      . . .
9  };
```

A complex number consists of `real` and `image` parts. In this case, `real` and `image` are data members of the class. Also, the method `module` is defined to calculate the module of a complex number.

In C++, a structure is a class defined by the keyword `struct`. The main difference between structures and classes is that all members of a structure by default are public.

2.2.2 Methods

The **methods of a class** can be inside or outside a class declaration. In the first case, a compiler tries to create an inline function whose code will be placed in the point of a method call. If method calls are time-consuming, inline functions can increase execution efficiency by increasing the code size.

For instance, the calculation of a complex number argument can be implemented in the class `Complex` as an inline function.

Listing 2.7.

```
1  #define Pi 3.1415926
2  . . .
3  // Argument of a complex number
4  double argument() {
5      return atan2(image, real) * 180 / Pi;
6  }
```

A definition of methods outside a class declaration does not differ from a definition of usual functions except for the requirement that the name of the class and the scope resolution operator : : must be before the name of the function.

Listing 2.8.

```
1  // Complex number module
2  double Complex::module() {
3      return sqrt(real * real + image * image);
4  }
```

2.2.3 Objects of a class

After a definition of a class and its behavior, the class name is used as the name of any data type; for example, int, float. Complex numbers can be defined as follows:

Listing 2.9.

```
1  // Complex number
2  Complex c1, c2, c3;
```

The declared variables c1, c2, c3 are the **objects of the class** Complex. An object supports the call of member functions of their class by using the operator . (dot).

Listing 2.10.

```
1  double a, b;
2  a = c1.module();
3  b = c3.argument();
4  printf("a = %f, b = %f\n", a, b);
```

A pointer to an object can be defined in C++ as for any other variable. The operators new and delete are employed for dynamic creation and deletion of objects. If the access to an object is implemented by a pointer, the operator -> is used instead of the operator . (dot).

Listing 2.11.

```
1  Complex *p1 = new Complex;
2  printf("Argument = %f\n", p1->argument());
3  delete p1;
```

2.2.4 Constructors and destructors

After an object is created, all its data must be assigned values. In C++, the initialization and deletion of objects can be performed by constructors and destructors, respectively. A **constructor** is a special block of instructions called during object creation and designed to initialize an object. A constructor has the same name as the class and returns nothing. A class can have several constructors with different sets of arguments.

Listing 2.12.

```
1  // Default constructor
2  Complex() {
3  }
4  // The first constructor
5  Complex(double r) {
6      real = r;
7      image = 0;
8  }
9  // The second constructor
10 Complex(double r, double i) {
11     real = r, image = i;
12 }
```

If a constructor is not defined, a compiler generates a default constructor that has no parameter and does nothing. If a class has other constructors, a default constructor has to be defined directly.

A constructor is called when an object is created.

Listing 2.13.

```
1  c1 = Complex(5.0, 4.0);
2  c2 = Complex(3.0);
3  c3 = Complex();
```

A **destructor** is called every time an object is destroyed. The destructor has the same name as the constructor, but it is preceded by a ~. A destructor has no arguments and returns nothing.

Listing 2.14.

```
1  // Destructor
2  ~Complex() {
3  }
```

A compiler creates a destructor if it is not directly defined. When a function, where a local object was created, completes its work, a destructor of an object is automatically called. A direct call of the destructor is not required. If an object was created using new, then delete is used to call the destructor.

2.3 Function and operator overloading

Function overloading is the process of using the same name for two or more functions. Function and operator overloading illustrates the polymorphism of C++, i.e. *one interface, many methods.*

2.3.1 Function overloading

Several functions with the same name can be used in C++; each of them can perform different actions and must have different lists of parameters. This capability is convenient to use for performing actions with same meaning on objects of various types.

A simple example is the **overloaded function** for printing values with types int, double, and char*.

Listing 2.15.

```
1  #include <stdio.h>
2  void print(int i) {
3      printf("%d\n", i);
4  }
5  void print(double x) {
6      printf("%f\n", x);
7  }
8  void print(char* s) {
9      printf("%s\n", s);
10 }
11 int main() {
12     int n = 17;
13     float pi = 3.1415926;
14     print("An example of the print"); // Call print(char* s)
15     print(n); // Call print(int i)
16     print(pi); // Call print(double x)
17     return 0;
18 }
```

2.3.2 Operator overloading

In C++, overloading is also permitted for operators. Thus, operators can be used not only for built-in types, but also for classes.

Overloaded operators exist in the language itself. For example, the addition operator works with different types int, float, and double, and not only with them. Let @ be some operator of C++ (apart from . .* :: ?:). Then it is enough to define a function with the name operator@ and necessary arguments for this function to perform the necessary actions. These actions

```
a + b;
a.operator+(b);
```

are equivalent.

Overloading of the addition operator in the class of complex numbers is given as follows.

Listing 2.16.

```
1  class Complex {
2      double real, image;
3  public:
4      Complex(double r, double i = 0) {
5          real = r, image = i;
6      }
7      Complex operator+(const Complex &a) const;
8  };
9  // Overloading of addition operator
10 Complex Complex::operator+(const Complex &a) const {
11     return Complex(real + a.real, image + a.image);
12 }
```

In this example, the keyword const provides immutability of the argument and object.

2.3.3 Operator overloading with friend functions

Access specifiers directly indicate whether functions outside of the defined class can access its members or not. The friendship concept provides for a certain function or class the access to elements of a class, which are specified as private or protected. To declare a function as a **friend** of a class, the keyword friend is put before the function prototype in the class definition.

In the following we use friend functions for overloading the multiplication operator in the work with complex numbers.

Listing 2.17.

```
1  class Complex {
2      double real, image;
3  public:
4      Complex(double r, double i = 0) {
5          real = r, image = i;
6      }
7      friend Complex operator*(const Complex &a, const Complex &b);
8  };
9  // Multiplication operator overloading
10 Complex Complex::operator*(const Complex &a) const {
11     return Complex(a.real * b.real - a.image * b.image,
12         a.real * b.image + a.image * b.real);
13 }
```

2.4 Inheritance

The reuse of high-quality software saves time and effort in the development of new programs. Inheritance is considered to be the main approach to reuse software. New classes are created from already existing classes by taking their attributes and functions and adding new features.

2.4.1 Concept of inheritance

In the creation of a new class, instead of writing completely new data-members and methods we can declare that the new class is an inheritor of an earlier defined class (the base class). The new class is called a **derived class**. In turn, any derived class can serve as the base class for subsequent classes, i.e. we create hierarchical classifications. We speak of simple inheritance if a new class is derived from one base class, and multiple inheritance when a new class is derived by several base classes.

A derived class adds its own data members and functions, and, in this sense, a derived class is richer than the base class. On the other hand, a derived class is more narrow than its base class and operates with a smaller group of objects. Each object of a derived class is also an object of the corresponding base class.

The idea of inheriting existing collections of classes by new classes provides the basis for the hierarchical organization of current software. Individual collections of classes are developed using various widespread libraries available for use. Within the concept of component-based programming, software is designed from the standard repeatedly-used components.

2.4.2 Protected elements

In defining base classes, we may observe elements declared as public, private, and protected. A derived class can use public elements of the base class just as these elements are defined in the derived class. At the same time, the derived class cannot directly work with the private elements of the base class. To use such elements, the derived class must have some interface functions.

Protected elements of the base class stand at an intermediate position between private and public elements. If an element is protected, objects of the derived class can directly apply to the element, as it is public. Direct access to protected elements is impossible for other parts of a program.

2.4.3 Example: a point and circle

As an example of inheritance, we consider the class of points on a plane (the base class) and the class of circles (the derived class).

Listing 2.18.

```
1  // Point class
2  class Point {
3  protected:
4      float x, y; // point's coordinates
5  public:
6      Point(float a = 0, float b = 0);
7      void setPoint(float, float);
8      float getX() const { // get x coordinate
9          return x;
10     }
11     float getY() const { // get y coordinate
12         return y;
13     }
14 };
15 // Constructor by default
16 Point::Point(float a, float b) {
17     setPoint(a, b);
18 }
19 // set coordinates
20 void Point::setPoint(float a, float b) {
21     x = a;
22     y = b;
```

```
23  }
24  // Circle class
25  class Circle : public Point { //public inheritance
26  protected:
27      float radius; // circle radius
28  public:
29      Circle(float r = 0, float x = 0, float y = 0);
30
31      void setRadius(float r) { // set radius
32          radius = r;
33      }
34
35      float getRadius() const { // get radius
36          return radius;
37      }
38  };
39  // Constructor
40  Circle::Circle(float r, float a, float b) :
41  Point(a, b) // Base class constructor call
42  {
43      radius = r;
44  }
```

The class `Circle` can be used as the base for describing such geometric objects as a cone or cylinder. Thereby, the element `radius` is declared to be `protected`.

2.4.4 Virtual functions

A **virtual function** in C++ is a function that can be redefined in derived classes; a specific realization of this function for call is defined during execution.

Derived classes can redefine a virtual function of the base class or not redefine it. Types of arguments, their number and a returned value type of inherited class function must be the same as for the function of the base class. A virtual function is defined using the keyword `virtual` in a function declaration.

For example, the virtual function `print` can be used for the considered classes `Point` and `Circle`.

Listing 2.19.

```
1  // point class
2  . . .
3  public:
4  virtual void print() const;
5  . . .
6  // print function of point
```

```
7   void Point::print() const {
8       printf("Point = (%f,%f)\n", x, y);
9       . . .
10      // circle class
11      . . .
12  public:
13      virtual void print() const;
14      . . .
15  // print function of circle
16  void Circle::print() const {
17      printf("Circle: radius = %f, center = (%f,%f)\n", radius, x, y);
18  }
```

Listing 2.20.

```
1   #include <stdio.h>
2   int main() {
3       Point* p;
4       p = new Point(0.4, 0.3);
5       p->print();
6       delete p;
7       p = new Circle(0.1, 0.5, 0.2);
8       p->print();
9       delete p;
10      return 0;
11  }
```

The output of the program is

```
Point = (0.400000,0.300000)
Circle: radius = 0.100000, center = (0.500000,0.200000)
```

Virtual functions realize such an important feature of C++ as polymorphism. Polymorphism is the ability for objects of different classes which are associated by inheritance to react differently while calling the same function.

2.5 Input/output

A part of the C++ standard library is the **iostream library,** which provides a set of features to perform input/output operations. This library supports input/output of data from/to files for built-in data types. Within the concept of object-oriented programming, using one's own classes, it is possible to extend this library in order to read and write new data types.

2.5.1 Output to streams

To use the iostream library, it is necessary to include the header file

```
#include <iostream>
```

The support of input operations is provided by the class istream; the support of output operations is provided by the class ostream. The derived class of these base classes of iostream supports bidirectional input/output.

In the library iostream, the following streams are defined:

- cin – the object of the class istream that corresponds to the standard input (keyboard);
- cout – the object of the class ostream that performs the standard output (display);
- cerr – the object of the class ostream that implements the standard output of errors.

The class ostream ensures, first of all, input of data of the standard types using the overloaded operator <<. This operator is overloaded to output data with built-in types and strings.

Listing 2.21.

```
1  #include <iostream>
2  using namespace std;
3  int main() {
4      char *s = "Test";
5      int m = 7;
6      float pi = 3.14159;
7      cout << "Example:\n    m = " << m
8          << ", 2*pi = " << 2 * pi
9          << ", s = " << s << endl;
10     return 0;
11 }
```

The output of the program is

```
Example:
    m = 7, 2*pi = 6.28318, s = Test
```

In this example, the overloaded operator << is employed in the concatenated form. The pass to the new line is performed by using both the escape sequence \n and the manipulator endl (the end of a line). The manipulator endl leads also to an output buffer reset.

The namespace std is used in the C++ standard library (in our example, using namespace std;). omit the direct declaration of the namespace, std::cout and std::endl need to be used instead of cout and endl, respectively.

2.5.2 Input into a stream

The overloaded operator >> is employed to input data. The program to output the multiplication of two entered simple numbers is given below.

Listing 2.22.

```
1  #include <iostream>
2  using namespace std;
3  int main() {
4      int n, m;
5      cout << "Input n and m:\n";
6      cin >> n >> m;
7      cout << "Result: n*m = " << n * m << endl;
8      return 0;
9  }
```

Output of the program execution:

```
Input n and m:
3
4
Result: n*m = 12
```

2.5.3 Overloading of input/output operators

If we want our class to support input/output operations, then the operators << and >> need to be overloaded. Input and output of a complex number on the basis of overloading of << and >> is shown below.

Listing 2.23.

```
1  #include <iostream>
2  using namespace std;
3  class Complex {
4      double real, image;
5  public:
6      Complex() {
7      }; // Constructor by default
8      Complex(double r) {
9          real = r;
10         image = 0;
11     } // Constructor
12     ~Complex() {
13     } // Destructor
14     friend ostream& operator<<(ostream &fo, Complex &c);
15     friend istream& operator>>(istream &fi, Complex &c);
16 };
17 // Overloaded operator <<
18 ostream& operator<<(ostream &fo, Complex &c) {
19     if (c.image > 0)
20         fo << c.real << "+i" << c.image << "\n";
21     else
22         fo << c.real << "-i" << -c.image << "\n";
23     return fo;
24 }
25 // Overloaded operator >>
26 istream& operator>>(istream &fi, Complex &c) {
27     cout << "Input of real part: ";
28     fi >> c.real;
29     cout << "Input of image part: ";
30     fi >> c.image;
31     return fi;
32 }
33 int main() {
34     Complex c;
35     // Input a complex number
36     cin >> c;
37     // Output the complex number
38     cout << "c = " << c << endl;
39     return 0;
40 }
```

The output is as follows:

```
Input of real part: 1
Input of image part: -2
c = 1-i2
```

2.5.4 Formatted input/output

Problems of formatting in C++ are resolved by the functions setf, unsetf, and flags of the class ios. The second possibility is associated with the use of manipulators (the header file iomanip). We can assign a width of input/output fields, output data with alignment, output integer numbers in decimal, octal, and hexadecimal forms, output floating numbers with different precision, and so on.

The function flags must assign the value that installs all flags (the type long). The function setf is intended to change one or more flags (the operation | is employed to repeat) for the current state of the format. The function unsetf is used to reset a flag (flags). Similar problems are resolved by applying manipulators setiosflags and resetiosflags.

Listing 2.24.

```
1  #include <iostream>
2  #include <iomanip>
3  using namespace std;
4  int main() {
5      int n = 123456789;
6      double a = 734.5831700;
7      cout << "n = " << n << endl; // standard output
8      cout << setw(20) << n << endl; // width of output
9      // hexadecimal format
10     cout << setbase(16) << n << endl;
11     cout << setbase(10) << setw(15) << setiosflags(ios::right)
12         << n << endl; // right-aligning
13     cout << "a = " << a << endl; // standard output
14     cout << setprecision(12) << a << endl; // precision of
15     // exponential format
16     cout << setiosflags(ios::scientific) << a << endl;
17     return 0;
18  }
```

The output of the program is as follows:

```
n = 123456789
    123456789
    75bcd15
    123456789
a = 734.583
    734.58317
    7.345831700000e+002
```

2.5.5 Reading/writing from/into files

To work with files in C++, it is necessary to include the header file `fstream`:

```
#include <fstream>
```

To read data from a file, objects of the class **ifstream** are used; to write data into a file, objects of the class **ofstream** is applied. The class `fstream` is employed to read and write. Files are opened by creating objects of these classes. A constructor of an object has two arguments: a file name and an opening mode. For instance, for the object `ofstream`, the opening mode can be `ios::out` to output data without modification to a file.

Table 2.1. Opening mode.

Parameter	Mode
ios::app	Write data to the end of the file
ios::ate	Shift to the end of the opened file
ios::in	Open the file for input
ios::out	Open the file for output
ios::trunc	If the file exists, then it will be cleaned
ios::nocreate	If the file does not exist, then a new file will not be created
ios::noreplace	If the file exists, then it will not be opened

The function **open** is used to link a stream to a file (to open the file for reading or writing). After the file is opened and linked to the stream, the work with the file is performed in the same way as with the standard input/output streams `cin` and `cout`.

When the end of the file EOF is reached, nothing can be read. To check a file state, we call the function **eof()**, which returns `true` if the end of the file is reached, and `false` otherwise. The state of a file stream can be checked by using the stream identifier as a logical condition. After finishing the input/output operation, the file must be closed by the function **close()**.

Listing 2.25.

```
1  #include <iostream>
2  #include <fstream>
3  #include <iomanip>
4  using namespace std;
5  int main() {
6      int n;
7      float a;
8      ifstream in("Test_in.dat");
```

```
9     if (!in) {
10    cout << "Error opening input file.\n";
11    return -1;
12    }
13    n = 0;
14    while (in >> a) {
15    n++;
16    cout << setw(10) << setiosflags(ios::left) << n
17        << setiosflags(ios::scientific) << a << endl;
18    }
19    return 0;
20    }
```

If the contents of the file `Test_in.dat` is

```
5.826622 1.106405 1.932502 0.071158 2.978757
```

then the result of the program execution is

```
1    5.826622e+000
2    1.106405e+000
3    1.932502e+000
4    7.115800e-002
5    2.978757e+000
```

2.6 Standard libraries

The **standard library of C++** provides a collection of common classes and interfaces which expand significantly the core of C++. Some components of this library, which can be useful in a numerical solution of applied problems, are highlighted below. In most cases, the use of the C standard library is appropriate.

2.6.1 Work with time

The efficiency of a numerical algorithm is estimated, in particular, by time expenses. To work with time, the header file `time.h` of the C standard library or **ctime** of the C++ standard library should be included.

The **function time(NULL)** gets the current time `time_t`, with the value in seconds counted from January 1, 1970. The **function ctime** returns time as a string: weekday, month, day, time (hour, minutes, seconds) and year.

The **function clock** returns time (the type clock_t) measured in processor cycles from the beginning of the program execution. The constant CLOCKS_PER_SEC determines the number of processor cycles per second.

Listing 2.26.

```
1   #include <iostream>
2   #include <ctime>
3   using namespace std;
4   int main() {
5       time_t lt;
6       clock_t start = clock(), finish;
7       double x = 2., e1 = 0.5;
8       lt = time(NULL);
9       cout << ctime(&lt) << endl;
10      for (int i = 3; i <= 100000; i++) { // calculating the number e
11          x = -x * i;
12          e1 = e1 + 1 / x;
13      }
14      finish = clock();
15      cout << "e =" << 1 / e1 << ",\n" << "CPU time: "
16          << 1000 * (finish - start) / CLOCKS_PER_SEC << " ms.\n";
17      return 0;
18  }
```

2.6.2 Common functions

We now need to discuss some useful utilities of the header file cstdlib of the C++ standard library (stdlib.h in C).

We have already met with some examples of common functions. For instance, we discussed above dynamic memory control using the library functions malloc, calloc, and free. Functions of search and sort can be of interest (bsearch and qsort).

We also note the possibility of transforming types. The **functions atof** and **atoi** are used to convert a string to a floating number and integer, respectively.

The following functions are intended to control the execution of a program: exit terminates the program, abort incorrect termination of the program, system executes an external command (command line).

Listing 2.27.

```
1   #include <iostream>
2   #include <cstdlib>
3   using namespace std;
4   int main() {
5       int n1, n2, n3;
6       char s1[10], s2[10], s3[10];
7       cout << "Input nonzero n1: ";
8       cin >> s1;
9       n1 = atof(s1);
10      cout << "Input nonzero n2: ";
11      cin >> s2;
12      n2 = atof(s2);
13      cout << "Input n1 x n2: ";
14      n3 = atof(s3);
15      cin >> s3;
16      n3 = atof(s3);
17      if (n1 == 0 || n2 == 0 || n3 == 0) {
18          cout << "Invalid input!" << endl;
19          exit(-2);
20      }
21      if (n3 != n1 * n2) {
22          cout << n3 << "!=" << n1 << "x" << n2 << endl;
23          system("calc"); // calculator
24          exit(-1);
25      }
26      cout << "OK: " << n3 << "=" << n1 << "x" << n2 << endl;
27      return 0;
28  }
```

2.6.3 Work with strings

In C, strings are considered as arrays of symbols. As for C++, a special class of the standard library is used to easily work with strings. The **class string** is located in the namespace std. The header file string must be included in order to use the class.

Let us discuss some opportunities for working with strings using the class string. The overloaded operator + can be employed to concatenate (add) strings. The operator [ ] or the function at can be applied to select a symbol in a specified place of a string. The operations of assignment and comparison are supported by the operators =, ==, !=. We can use the functions length (to get the length of a string), substr (to select a part of a string), find or rfind, copy, and replace. The function getline makes it possible to read a string from a specified stream.

Listing 2.28.

```cpp
1  #include <iostream>
2  #include <string>
3  using namespace std;
4  int main() {
5      string s1 = "Cauchy-Schwarz";
6      string s2 = "inequality";
7      s1 = s1 + " " + s2;
8      s2 = "Schwarz";
9      cout << s1 << endl;
10     size_t found;
11     found = s1.find(s2);
12     s1.replace(found, s2.length(), "Bunyakovskii");
13     cout << s1 << endl;
14     return 0;
15 }
```

Output of the program execution:

```
Cauchy-Schwarz inequality
Cauchy-Bunyakovskii inequality
```

Victor S. Borisov and Petr N. Vabishchevich

3 GCC distilled

Abstract: The **GNU Compiler Collection (GCC)**[1] is a collection of compilers for various programming languages. GCC has been developed within the GNU Project[2]. Here we discuss how to employ GCC in order to compile, link, debug, and organize C and C++ programs.

3.1 General information

GCC is a free software and the standard compiler for Linux and other UNIX-like operating systems.

3.1.1 Supported languages

The compilation process is divided into two phases. During the first phase (the front end), a compiler analyzes a source code and converts it into internal instructions in the form of an abstract tree that is independent from languages and processors. During the second phase (the back end), these instructions are processed to create a code working on a given platform.

At present, front ends are written for various programming languages, and back ends are designed for all basic processors. GCC supports several basic programming languages: C, C++, Objective-C, Java, Fortran, Ada, and Go.

Historically, the first meaning of the GCC abbreviation was associated with GNU C Compiler. In this case, the compilation of C programs is emphasized.

GCC is written basically in C. The distribution kit contains standard libraries for C++, Java, and Ada,

3.1.2 Platforms

The GCC collection works on many platforms. A platform is a combination of a certain processor and operating system.

In fact, GCC is the standard compiler for Linux systems. Note that GCC is compatible with many platforms. Several basic platforms are used to test the correct work

---

1 http://gcc.gnu.org/.
2 http://www.gnu.org/.

of release versions of this software. The basic platforms are the following operating systems: Debian Linux, Red Hat Linux, and FreeBSD on Intel x86-64. Significant attention is also given to porting GCC to Windows.

3.1.3 Installation on Linux

For most users, the easiest way to install GCC is to install a package designed for your operating system[3]. The GCC Project does not provide binary files and contains only source codes.

The good news for Linux users is that all GNU/Linux distribution kits contain GCC. The user may only need to expand the list of installed components. For instance, one can install the GCC documentation.

Programs on **Ubuntu** are installed through the internet by downloading them from repositories. The installation is performed via a graphical user interface (GUI) or console program.

The simplest way to install a new program on Ubuntu is to employ the **Ubuntu Software Center** in the Launcher. One can also use the Software Center to remove programs and search for a program by name or description.

The **Synaptic Package Manager** provides advanced package management capabilities for Ubuntu. This program combines all the features of the command-line package manager apt with a convenient graphic interface. Employing Synaptic, we can install, remove, configure, and upgrade packages for our system, browse among the list of available and installed packages, manage repositories, and upgrade the operating system.

The **Advanced packaging tool (apt)** can be used to manage packages on Ubuntu. We can install programs applying apt (apt-get). To install a new program xxx, the command apt-get install xxx is used. Downloading, installation, and configuration of programs are performed automatically. If the configuration needs more information, a request to the user will be displayed. The command apt-get remove xxx is utilized to remove the program xxx.

3.1.4 Work with GCC on Windows

The user can employ a precompiled and ready-to-work version of GCC on a **Windows** operating system in the **Cygwin** Project. Cygwin provides the integration of Windows-based applications, data, and resources with a UNIX-like environment. Cygwin is a powerful and advanced free collection of tools for porting UNIX programs to Windows

---

3 http://gcc.gnu.org/install/.

and cross-compilation (creating a binary code for one platform on other one). Cygwin has a large collection of applications which ensure a normal UNIX environment. In particular, besides GCC, Cygwin contains many other GNU development tools to perform basic programming tasks.

On the official website of Cygwin[4], we can find the information about the current status of the project, software updates, and a list of ftp-servers from where Cygwin packages can be installed. The setup.exe file is used to install Cygwin. Applying it, only those components which are really needed can be selected. On the page Select Packages, packages are selected for downloadibg. In the development tools category Devel, the following packages should be selected: gcc-core (the C compiler), gcc-g++ (the C++ compiler), gdb (GNU debugger), and make (a tool to work with makefiles). For convenient application other packages may also be needed.

The **Minimalist GNU for Windows (MinGW)** Project was separated from Cygwin some years ago. MinGW[5] Project provides GCC compilers with a minimal set of development tools for Windows applications. Cygwin provides much more complete compatibility with UNIX. The Minimal SYStem (MSYS) component provides a small independent environment with a lightweight UNIX-like shell. MinGW binaries and its components are available ans free to download[6].

The main difference between MinGW and Cygwin is that in MinGW executive files are free from any additional dependencies. We also point out that there is a difference in licensing. Cygwin is distributed under the terms of the GPL license, which can impose certain restrictions on the commercial use of a developed software. MinGW is more liberal from this point of view and can be more appropriate for commercial use.

3.2 Available documentation

Standard methods for getting help with GCC using the utility man are available. Work with the most complete and flexible system of software documentation info is discussed below.

3.2.1 Documentation on the internet

A workable system implies the possibility of obtaining help on various questions. Questions may be different, which means different types of help information and ways to get them.

---

4 http://cygwin.com/.
5 http://www.mingw.org/.
6 http://sourceforge.net/projects/mingw/files/MinGW/.

For GCC, we can use resources on the internet. We can begin with pages of online documentation of GCC[7]. Here we mention, in particular, Using the GNU Compiler Collection.

As in other modern projects, on the website of GCC, the user can find GCC wiki[8] or get answers to frequently asked questions (FAQs)[9].

Among the available online tutorials, An Introduction to GCC for the GNU Compilers gcc and g++[10] by Brian Gough should be mentioned as very useful.

3.2.2 Help via –help

Let us consider the basic ways of getting help with GCC and other programs on a Linux system. On Windows systems, access to help information is provided by using Cygwin. In this regard, MinGW is not as good, the user will find little information.

The **quick reference** for most Linux programs is easy to call up by running a program with the option --help (or -h). For instance, for the C++ compiler (the command g++), we have

```
$ g++ --help
Usage: g++ [option] file...
Options:
  -pass-exit-codes       Exit with the highest error code from the pass
  --help                 Display this information
  --target-help          Display platform specific command-line options
  --help={target|optimizers|warnings|params|[^]{joined|separate|
undocumented}}[,...]
                         Display specific types of command-line options
  (Use '-v --help' to display command line options of sub-processes)
  --version              Display compiler version information
. . .
```

To display the pages, the command more is used in combination with --help|more. After the screen is full, the command pauses and displays in the bottom line the message --More--. By pressing Enter, the next line of the text is displayed. By pressing SPACE, the next full screen will appear.

---

7 http://gcc.gnu.org/onlinedocs/.
8 http://gcc.gnu.org/wiki/.
9 http://gcc.gnu.org/faq.html.
10 http://www.network-theory.co.uk/docs/gccintro/index.html

The command less provides more functionality: navigation through a page using back and forward, scrolling a text with arrows, transition to a specific line of the text, search in the text, etc.

3.2.3 Manual pages

The standard reference documentation on Linux is executed as **man (manual) pages**. To access manual pages, we run the man command with the name of a program. An example for man gcc is

```
GCC(1)                          GNU                          GCC(1)


NAME
     gcc - GNU project C and C++ compiler


SYNOPSIS
     gcc [-c|-S|-E] [-std=standard]
         [-g] [-pg] [-Olevel]
         [-Wwarn...] [-pedantic]
         [-Idir...] [-Ldir...]
         [-Dmacro[=defn]...] [-Umacro]
         [-foption...] [-mmachine-option...]
         [-o outfile] [@file] infile...


     Only the most useful options are listed here; see below for the
     remainder.  g++ accepts mostly the same options as gcc.


DESCRIPTION
     When you invoke GCC, it normally does preprocessing, compilation,
     assembly and linking:  The "overall options" allow you to stop
     this process at an intermediate stage.  For example, the -c option
     says not to run the linker. Then the output consists of object
     files output by the assembler.
. . .
```

Manual pages for Linux are split into sections which reflect the functionality of programs (see Table 3.1). The man program looks for manual pages just in this order. Manual pages are browsed with the command less.

The structure of manual pages is well-established. Usually, they contain the following parts: NAME – the name and a brief description; SYNOPSIS – how to use the command; DESCRIPTION – the description of the command functionality in detail; EXAMPLES – examples of use; SEE ALSO – related man pages.

The command apropos is used for searching by keywords in the names of man pages and short descriptions of manual pages. The command whatis is applied to

Table 3.1. Manual page section.

Section	Description
1	Executable programs or shell commands
2	System calls
3	Library calls
4	Special files
5	File formats
6	Games
7	Miscellanea
8	System administration commands

search for man pages. It searches in the names of man pages and displays a brief description of found pages.

```
$ apropos compiler
c++ (1)                 - GNU project C and C++ compiler
c89 (1)                 - ANSI (1989) C compiler
c89-gcc (1)             - ANSI (1989) C compiler
c99 (1)                 - ANSI (1999) C compiler
c99-gcc (1)             - ANSI (1999) C compiler
cc (1)                  - GNU project C and C++ compiler
f95 (1)                 - GNU Fortran compiler
g++ (1)                 - GNU project C and C++ compiler
g++-4.6 (1)             - GNU project C and C++ compiler
gcc (1)                 - GNU project C and C++ compiler
gcc-4.6 (1)             - GNU project C and C++ compiler
gfortran (1)            - GNU Fortran compiler
gfortran-4.6 (1)        - GNU Fortran compiler
. . .
```

3.2.4 Use of info

The advanced help system for Linux programs is known as **info pages**. In contrast to man pages, info supports hypertext, which makes it possible to easily move around a document.

The list of existing info pages is available by running the info command:

```
File: dir,      Node: Top       This is the top of the INFO tree

  This (the Directory node) gives a menu of major topics.
  Typing "q" exits, "?" lists all Info commands, "d" returns here,
  "h" gives a primer for first-timers,
```

```
"mEmacs<Return>" visits the Emacs manual, etc.

In Emacs, you can click mouse button 2 on a menu item or cross
reference to select it.

* Menu:

Archiving
* Cpio: (cpio).          Copy-in-copy-out archiver to tape or disk.
. . .
```

To get help on a command, `info` with the name of the command must be executed. We can move around the document using the arrow keys: page navigation is performed via `Page Up` and `Page Down`. An example of the command `info gcc`:

```
File: gccgo.info,  Node: Top,  Next: Copying,  Up: (dir)

Introduction
************

This manual describes how to use 'gccgo', the GNU compiler for the Go
programming language.  This manual is specifically about 'gccgo'.  For
more information about the Go programming language in general,
including language specifications and standard package documentation,
see 'http://golang.org/'.

* Menu:

* Copying::                       The GNU General Public License.
* GNU Free Documentation License::
                                  How you can share and copy this manual.
* Invoking gccgo::                How to run gccgo.
* Import and Export::             Importing and exporting package data.
* C Interoperability::            Calling C from Go and vice-versa.
* Index::                         Index.
. . .
```

3.2.5 Tools to visualize help information

The above-considered `man` and `info` pages are focused on the command-line interface. This may be inconvenient to use.

There are special programs which permit viewing of `man` pages in graphics mode. `Xman` should be mentioned among them.

Konqueror – a web-browser, file manager and universal document viewer – provides ample functionality for viewing and navigation through help documentation.

The **yelp** program demonstrates a similar functionality.

To view man pages of the ls command, we can run yelp man:ls (or yelp info:ls) (see Figure 3.1). In the Konqueror browser, similar commands info:gcc (or man:gcc) are employed (see Figure 3.2).

3.2.6 Additional help

Most man pages are located in the directory /usr/share/man. The manpath command returns the full list of search directories for manual pages. The files of the info pages are located in the directory /usr/share/info.

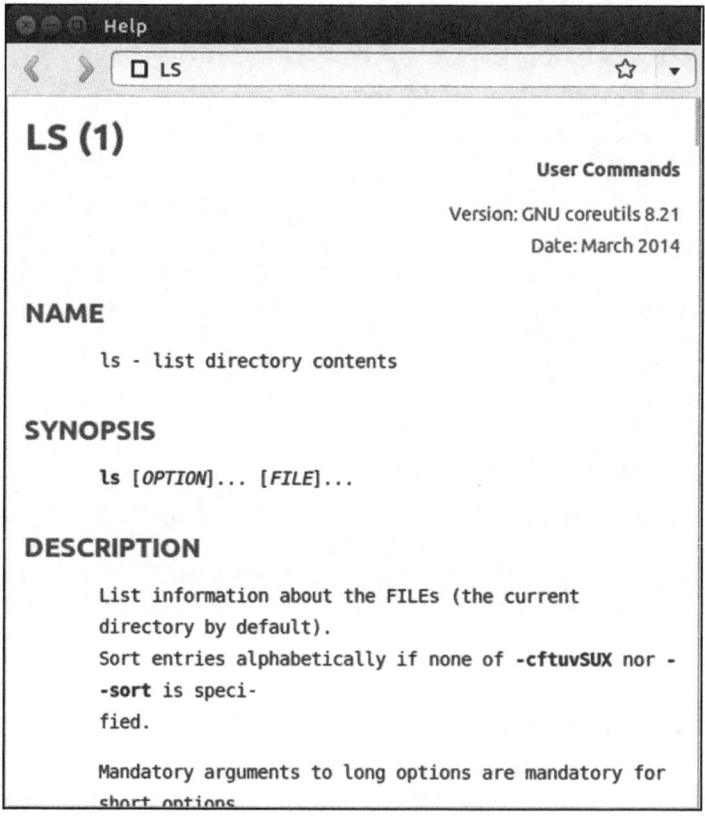

Fig. 3.1. Help in Yelp.

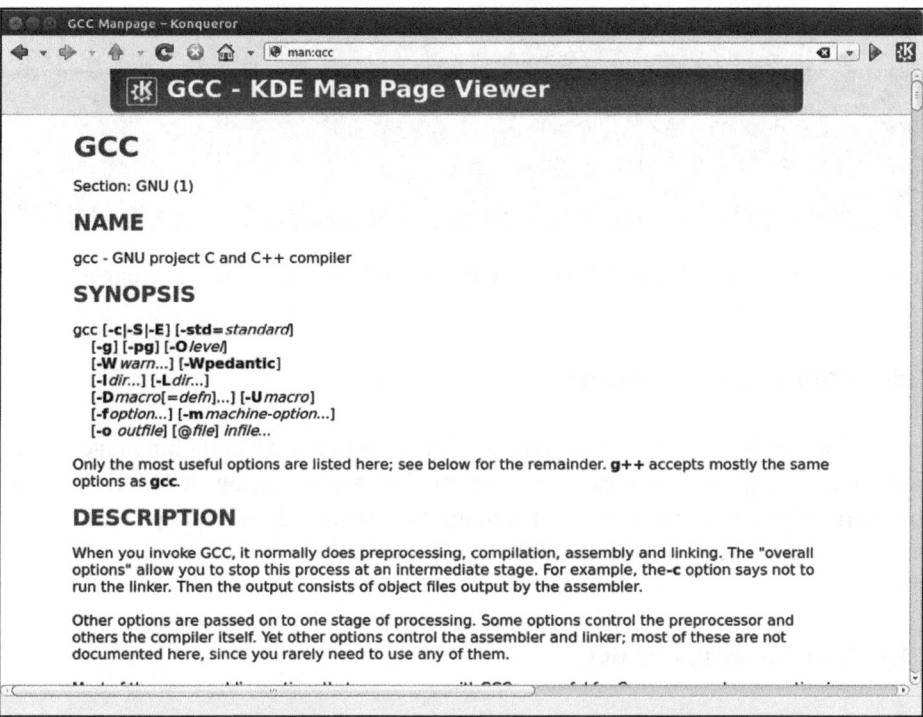

Fig. 3.2. Help in `Konqueror`.

For most programs, additional information is presented in other formats (`text`, `PDF`, `PostScript`, `HTML`) and located in the directory `/usr/share/doc`. An example is the contents of the directory `/usr/share/doc/gcc-4.6` (the `ls` command):

```
$ ls -l
total 6768
drwxr-xr-x 2 root root    4096 2011-10-19 19:39 C++
-rw-r--r-- 1 root root    1884 2011-09-17 01:34 changelog.Debian.gz
-rw-r--r-- 1 root root  133535 2011-09-16 16:47 changelog.gz
-rw-r--r-- 1 root root   25503 2011-09-17 01:18 copyright
drwxr-xr-x 2 root root    4096 2012-01-12 16:22 fortran
drwxr-xr-x 2 root root    4096 2011-10-12 18:28 gcc
-rw-r--r-- 1 root root 3070025 2011-09-17 01:20 gcc.html
-rw-r--r-- 1 root root 3399349 2011-09-17 01:20 gccint.html
drwxr-xr-x 2 root root    4096 2011-10-12 18:28 gomp
-rw-r--r-- 1 root root  149085 2011-09-17 01:20 libgomp.html
-rw-r--r-- 1 root root   15872 2011-09-16 16:46 NEWS.gz
-rw-r--r-- 1 root root   55729 2011-09-16 16:46 NEWS.html
drwxr-xr-x 2 root root    4096 2011-10-12 18:28 quadmath
-rw-r--r-- 1 root root   11158 2011-09-16 18:26 README.Bugs
```

```
-rw-r--r-- 1 root root    1509 2011-09-16 16:48 README.Debian
-rw-r--r-- 1 root root    2749 2011-09-16 16:47 README.Debian.amd64.gz
-rw-r--r-- 1 root root    2707 2011-09-16 16:50 README.Debian.i386.gz
-rw-r--r-- 1 root root    1165 2011-09-16 16:46 README.ssp
drwxr-xr-x 2 root root    4096 2011-10-19 19:39 test-summaries
-rw-r--r-- 1 root root    5967 2011-09-16 18:26 test-summary.gz
-rw-r--r-- 1 root root    1716 2011-09-16 16:48 TODO.Debian
```

In this way, we can find help information unavailable on man and info pages.

3.3 Compilation workflow

A **compiler** reads a set of instructions in a high-level programming language implemented as a program or separate program module and translates it into a set of instructions in machine language or a language close to machine language. To illustrate
the work of a compiler, let us consider a C++ program.

3.3.1 Up-and-running with GCC

The main stages of compiling a C program are presented in Figure 3.3.

The preprocessor is used by the compiler before processing a program to make
some changes in it. The preprocessor modifies the source code, includes additional
files with function definitions, replaces macros (reductions for fragments of the source
code) by the corresponding macros definitions, etc. Special preprocessor directives of
conditional compilation can be applied to include or exclude parts of a program. In
GCC, the executable program of the preprocessor is named cpp.

One of the main steps of compilation is the translation of a program or separate
program module written in C/C++ programming languages into assembler language
(as a file with .s extension), which is a low-level programming language. In GCC,
translation of a program is provided by the **program gcc** with the -S option.

The translation of a program into an executable machine code is performed by the
assembler. The assembler of GCC creates object files with the .o extension.

The final step is linking the executable file to the object files, a part of which may
be in the object libraries. For this, in GCC, the ld program is used. The complete program (a.out) can be executed on a computer.

3.3.2 Test program

We now illustrate the work of compilers on a test program written in the C++ language.

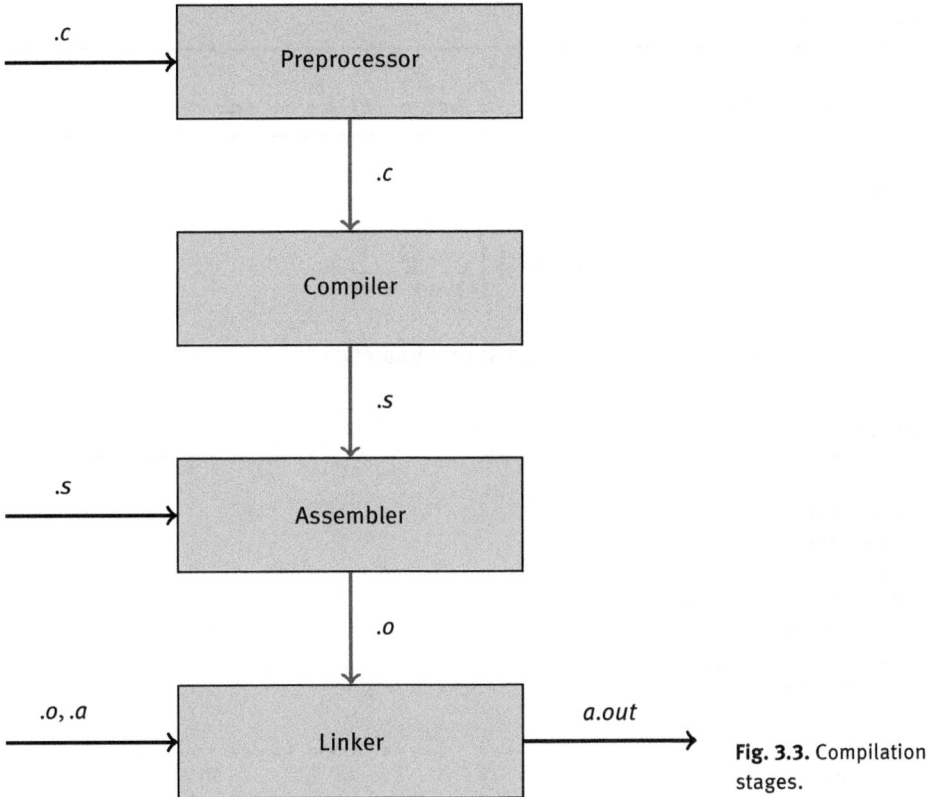

Fig. 3.3. Compilation stages.

The program consists of three files: int.cpp, int.h, and test.cpp. The intm function implements the computation of an integral by the rectangle method (the midpoint rule).

Listing 3.1.

```
1  // File: int.cpp
2  // Calculate an integral on an interval [a,b]
3  float intm(float(*func)(float), float a, float b, int n) {
4      float s = 0;
5      float h = (b - a) / n;
6      float x = a - h / 2;
7      for (int i = 0; i < n; i++) {
8          x += h;
9          s += func(x);
10     }
11     return s * h;
12 }
```

Listing 3.2.

```
1  // File: int.h
2  float intm(float(*func)(float), float a, float b, int n);
```

In `test.cpp`, the integral

$$I = 3\sqrt{3} \int_0^1 \frac{dx}{1 + x + x^2}$$

with the exact value $I = \pi$ is calculated approximately.

Listing 3.3.

```
1  // File: test.cpp
2  #include <iostream>
3  #include <math.h>
4  #include "int.h"
5  #define PI 3.141592653589
6  using namespace std;
7  float func(float x) {
8      return 3. * sqrt(3.) / (1. + x + x * x);
9  }
10 int main() {
11     for (int n = 1; n < 6; n++) {
12         float y = intm(func, 0., 1., 25 * n);
13         cout << "n = " << n << ", int = " << y - PI << "\n";
14     }
15     return 0;
16 }
```

3.3.3 Preprocessing

For the C++ language, we invoke the g++ compiler from GCC. To see the output of the preprocessor, we **run g++** with the -E option (or the cpp command). If we apply this option, the compiler stops after preprocessing. In the result, we obtain the source file including the contents of header files.

```
$ g++ -E test.cpp
# 1 "test.cpp"
# 1 "<built-in>"
# 1 "<command-line>"
# 1 "test.cpp"
```

```
# 1 "/usr/include/c++/4.6/iostream" 1 3
# 37 "/usr/include/c++/4.6/iostream" 3
. . .
extern int matherr (struct __exception *__exc) throw ();
# 476 "/usr/include/math.h" 3 4

# 4 "test.cpp" 2
# 1 "int.h" 1

float intm(float(*func)(float), float a, float b, int n);
# 5 "test.cpp" 2

using namespace std;
float func(float x) {
 return 3. * sqrt(3.) / (1. + x + x * x);
}
int main() {
 for (int n = 1; n < 6; n++) {
  float y = intm(func, 0., 1., 25 * n);
  cout << "n = " << n << ", int = " << y - 3.141592653589 << "\n";
 }
 return 0;
}
```

The `iostream` header file is included into the code. In turn, `iostream` comprises several header files associated with input and output operations. Further, the `math.h` header file contained the `sqrt` function, as well as our header file `int.h`, are also included in the code. In the main function, the string `PI` is replaced by `3.141592653589` according to the preprocessor directive.

3.3.4 Object code generation

The following command is employed to generate a code in the assembly language:

```
$ g++ -S test.cpp
```

In the result, the file `test.s` is created:

```
    .file   "test.cpp"
    .local  _ZStL8__ioinit
    .comm   _ZStL8__ioinit,1,1
    .text
    .globl  _Z4funcf
    .type   _Z4funcf, @function
_Z4funcf:
```

```
.LFB966:
    .cfi_startproc
    pushq   %rbp
    .cfi_def_cfa_offset 16
    .cfi_offset 6, -16
    movq    %rsp, %rbp
    .cfi_def_cfa_register 6
    movss   %xmm0, -4(%rbp)
. . .
```

The compiler command with the -c option is used to compile the test.cpp file.

```
$ g++ -c test.cpp
```

The output is a file test.o.

3.3.5 Executable file

Linking of object files into a single executable file test (a.out, by default) is performed by the following command:

```
$ g++ test.o int.o -o test
```

The output is as follows:

```
$ ./test
n = 1, int = -0.000231417
n = 2, int = -5.80867e-05
n = 3, int = -2.51849e-05
n = 4, int = -1.37409e-05
n = 5, int = -8.73406e-06
```

The complete sequence of preprocessing, compiling, and linking is implemented as follows:

```
$ g++ test.cpp int.cpp -o test
```

3.4 Compiling a C/C++ program

A brief description of commands and their options used to compile C/C++ programs is given in the following.

3.4.1 Input files for compiling

A compiler analyzes the names of the files passed to the compiler as arguments and determines what actions it has to perform. In particular, the compiler defines the language and includes the corresponding standard libraries. Tables 3.2 and 3.3 present a brief description of input files for C and C++, respectively.

We can explicitly instruct the compiler to consider source files with the suffix .c and .h as C++ source files. To specify the language, the gcc command with the -x LANGUAGE option is used. The LANGUAGE option is, e.g. c, c-header, cpp-output (for C) or c++, c++-header, c++-cpp-output (for C++). The second way is to invoke the g++ command instead of gcc (for C++).

Table 3.2. C files.

Extension	File description
.a	Static object library
.c	C source code
.h	Header file
.i	C source code that should not be preprocessed
.o	Object file
.s	Assembler code
.so	Shared object library

Table 3.3. C++ files.

Extension	File description
.a	Static object library
.C, .cc, .cpp, .c++, .cp, .cxx	C++ source code
.h, .hh, .hpp, h++	Header file
.ii	C++ source code that should not be preprocessed
.o	Object file
.s	Assembler code
.so	Shared object library

3.4.2 Options for searching for a directory

Directories for searching for header files and libraries are specified by options.

To get a list of directories to be searched by `gcc` for programs and libraries, we employ the command `gcc-print-search-dirs`. The output is

```
$ gcc -print-search-dirs
install: /usr/lib/gcc/x86_64-linux-gnu/4.8/
programs: =/usr/lib/gcc/x86_64-linux-gnu/4.8/
. . .
libraries: =/usr/lib/gcc/x86_64-linux-gnu/4.8/
. . .
```

The `-IDIR` option is applied to add the directory `DIR` to the beginning of the list of directories to be searched for header files. This allows changing of e.g. system header files, since system directories are scanned after. If we use several `-I` options, then the specified directories are scanned in left-to-right order.

The similar `-LDIR` option adds the directory `DIR` to a list of directories of libraries.

3.4.3 Options for controlling a language

A set of **options controls** dialects of the languages that the compiler accepts. The option `-std=STANDARD` is applied to define the language standard version `STANDARD`. For instance, `-std=C90` is used to determine the `C90` standard for the `C` language. The `GNU` dialect of the `C90` standard, which includes some features of `C99`, is selected using `-std=gnu90`.

For the `C++` language, the option `-std=c++98` corresponds to the `c++98` standard (`ISO/IEC 14882`), the `GNU` dialect of this standard is governed by `c++98`.

The `-ansi` option is equivalent to `-std=C90` and `-std=c++98` for C and C++, respectively.

3.4.4 Options for warnings

A set of options is available to handle **warnings**. A warning is a diagnostic message about constructions which are risky or probably contain an error.

The `-pedantic` option is used to print warnings about standard mismatches. In this case, the compliance with the language standard is checked and constructions forbidden by the standard version are rejected. This option is useful in order to achieve maximum portability.

Code syntax checking is provided by the -fsyntax-only option. The -Wall option is applied to print the main warnings. Also, the -Werror option transforms all warnings into errors.

3.4.5 Options for optimization

The GCC compiler supports various options for **optimization**. It is necessary to keep in mind that the impact of most of these options on the performance of a resulting program is ambiguous; in some cases, we get an increase in essential performance, whereas in other cases, a significant decrease in performance is observed.

The level of optimization is specified by the option -OLEVEL. The minimal optimization corresponds to LEVEL equals 1 (the option -O1), the moderate and aggressive levels are prescribed by 2 and 3, respectively.

At our own risk, we can use the -Os option to optimize (by reduction) the code size (based on -O2). If we consider -Ofast (the basic optimization -O3), the optimization is conducted by disregardibg the strict standard compliance.

3.5 Libraries and linking programs

A compiler creates object files which are linked with other object modules into an executable program. Here we will briefly discuss working with object modules, which can be implemented as static or dynamic libraries.

3.5.1 Simple linking

In the simplest case, object modules from separate files in a directory of a program are combined to form an executable program.

For our test program, using the commands

```
$ g++ -c int.cpp -o int.o
$ g++ -c test.cpp -o test.o
```

we compile the source files into the object files. Next, these object files are linked into the executable program test:

```
$ g++ int.o test.o -o test
```

3.5.2 Linking with a static library

A **static library** is a collection of object files created by the compiler. The names of static libraries usually begin with lib (prefix) and end with .a (suffix). The ar utility is used to work with the contents of a static library.

For example, after compilation of the file int.cpp via the command

```
$ g++ -c int.cpp
```

we obtain the object file int.o.

The ar command with the -r option is employed to create a new library from object files. In our case, we include the object file int.o into the library libint.a using the following command:

```
$ ar -r libint.a int.o
```

Further, we compile test.cpp with the static library libint.a:

```
$ g++ test.cpp libint.a -o test_s
```

A short form of library names can be used applying the option -l:

```
$ g++ test.cpp -L. -lint -o test_s
```

In the above case, we clearly indicate that the current directory contains the library.

The modules of a static library are included into a program only if they have functions and data called in the module that, in turn, is already used in linking. Therefore, linking with a static library generates a smaller executable file than linking with separate object files.

3.5.3 Linking with a shared library

A **shared (dynamic) library** is a collection of object files in which references to variables and function calls are relative rather than absolute. This allows loading and executing shared modules dynamically during the start and execution of a program. Here we have a small size of executable files, and moreover, several programs can simultaneously employ object codes from one shared library.

To build a shared library (suffix .so), it is necessary to prepare in a special way the object files which are to be included in the library. In our test program, we compile int.cpp by the following command:

```
$ g++ -c -fpic int.cpp
```

Special compilation into the position-independent code is provided by the -fpic option.

The next step is the creation of the shared library libint.so:

```
$ g++ -shared int.o -o libint.so
```

Compilation and creation of the shared library may be combined in the command

```
$ g++ -fpic -shared int.cpp -o libint.so
```

Further, we compile test.cpp using the shared library libint.so:

```
$ g++ test.cpp libint.so -o test_d
```

If we link a program with a static library, then all the object modules are located in a single executable file. This makes for portability of the program. In contrast, shared libraries must be available during linking and execution of the program, whereby any attempt to run the program test_d will be unsuccessful.

```
$ ./test_d
./test_d: error while loading shared libraries:
libint.so: cannot open shared object file:
No such file or directory
```

This message sends the dynamic linker, which cannot find the library libint.so.

The search for shared libraries is carried out in
- the directories mentioned in the environment variable LD_LIBRARY_PATH;
- the list of libraries from the file /etc/ld.so.conf;
- the directory /lib;
- the directory /usr/lib.

For example, we can add our current directory to the list /etc/ld.so.conf. After changes in the configuration file /etc/ld.so.conf, an update of the settings is provided by the command ldconfig.

3.6 Debugging

In the development of programs, the debugging stage is applied to identify and correct errors. The localization of errors is performed when a program is executed with the control of the current values of variables. In the following we demonstrate the usage of GNU Debugger (GDB)[11].

3.6.1 Compilation with GDB

GNU Debugger provides debugging of any applications. The full debugging is performed if the compilation and linking includes some debugging information about a source code of a program. Using a debugger, we can run any program line by line, study the variable values, and run a program until a certain prescribed point and stop at that point.

Special compiler options can be employed to set the level and type of debugging information. To generate minimal debugging information (tracing of function calls, global variables), we compile a program with the -g1 option (the first level). The second level, which corresponds to using the -g2 option (-g), generates the information about local variables and code lines. Full debugging information is available by using -g3.

It is not recommended to apply options for debugging together with options for optimization. The optimization complicates the program execution tracing.

We compile our test program with the debugging information as follows:

```
$ g++ -g int.cpp test.cpp -o test
```

Further manipulations are performed with the executable file test.

3.6.2 Getting started with GDB

To debug a program using GDB, we should define the name of the program as the first argument of gdb. We can also invoke the debugger and then load a program using the command load. In our example, we run GDB using the command gdb test. The output is as follows:

---

11 http://www.gnu.org/software/gdb/.

```
$ gdb test
GNU gdb (Ubuntu 7.7-0ubuntu3.1) 7.7
Copyright (C) 2014 Free Software Foundation, Inc.
License GPLv3+: GNU GPL version 3 or later
<http://gnu.org/licenses/gpl.html>
This is free software: you are free to change and redistribute it.
There is NO WARRANTY, to the extent permitted by law.
Type "show copying" and "show warranty" for details.
This GDB was configured as "x86_64-linux-gnu".
Type "show configuration" for configuration details.
For bug reporting instructions, please see:
<http://www.gnu.org/software/gdb/bugs/>.
Find the GDB manual and other documentation resources online at:
<http://www.gnu.org/software/gdb/documentation/>.
For help, type "help".
Type "apropos word" to search for commands related to "word"...
Reading symbols from test...(no debugging symbols found)...done.
(gdb)
```

Here (gdb) is a regular command prompt, which informs that the debugger is waiting for a new command.

3.6.3 Source code

The command list causes the source code of the program to be displayed:

```
(gdb) list
2    #include <iostream>
3    #include <math.h>
4    #include "int.h"
5    #define PI 3.141592653589
6    using namespace std;
7    float func(float x) {
8        return 3. * sqrt(3.) / (1. + x + x * x);
9    }
10   int main() {
11       for (int n = 1; n < 6; n++) {
(gdb)
```

Here ten lines of code are displayed. If the name of a function is defined in the command list, lines start from the beginning of the function. If the number of a line is given, then this line is printed first.

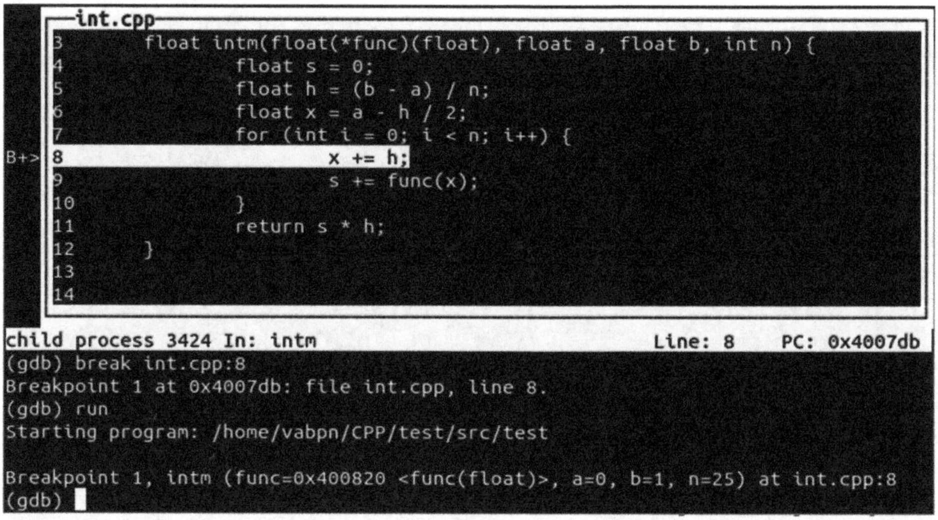

```
┌─int.cpp──────────────────────────────────────────────────────────────┐
│    3       float intm(float(*func)(float), float a, float b, int n) { │
│    4           float s = 0;                                            │
│    5           float h = (b - a) / n;                                  │
│    6           float x = a - h / 2;                                    │
│    7           for (int i = 0; i < n; i++) {                           │
│ B+> 8              x += h;                                             │
│    9               s += func(x);                                       │
│   10           }                                                      │
│   11           return s * h;                                          │
│   12       }                                                          │
│   13                                                                  │
│   14                                                                  │
└───────────────────────────────────────────────────────────────────────┘
child process 3424 In: intm                        Line: 8      PC: 0x4007db
(gdb) break int.cpp:8
Breakpoint 1 at 0x4007db: file int.cpp, line 8.
(gdb) run
Starting program: /home/vabpn/CPP/test/src/test

Breakpoint 1, intm (func=0x400820 <func(float)>, a=0, b=1, n=25) at int.cpp:8
(gdb) ▮
```

Fig. 3.4. Work of GDB text user interface.

The gdb with the -tui option (GDB text user interface) provides advanced features for debugging. The text interface has a separate text window to test a program. Figure 3.4 demonstrates an example of using TUI.

In TUI mode we can use the PgUp key to scroll a text one page up, and the PgDn key is employed to scroll one page down. The Up key makes it possible to move one line up, the Down key is applied to move one line down. The Left and Right keys are used to move left and right, respectively.

3.6.4 Breakpoints

The basic strategy for debugging by GDB is connected with the use of **breakpoints** for a running program and observation of internal data.

Using the break command, we can set a breakpoint referring to a function or line. If a program consists of several files, we define the name of a file with the : delimiter. For example, we set a breakpoint to the 8th line of the int.cpp file:

```
(gdb) break int.cpp:8
Breakpoint 1 at 0x4007db: file int.cpp, line 8.
(gdb)
```

The clear command removes breakpoints.

We can specify several breakpoints. The info breakpoints command displays the position and description of all breakpoints and indicate their number.

```
(gdb) info breakpoints
Num    Type          Disp Enb Address            What
1      breakpoint    keep y   0x00000000004007db in intm(float (*)(float),
float, float, int) at int.cpp:8
```

We can also enable or disable certain breakpoints by their numbers (enable, disable commands).

After setting breakpoints, we run the program using the run command:

```
(gdb) run
Starting program: /home/vabpn/CPP/test/src/test

Breakpoint 1, intm (func=0x400820 <func(float)>, a=0, b=1, n=25)
at int.cpp:8
8                         x += h;
(gdb)
```

Resuming execution of the program stopped by the debugger is performed by means of the continue command.

3.6.5 Displaying data

The debugger GDB allows us to easily check the data of a running program. The display command is used to display the values of indicated variables.

```
(gdb) display i
1: i = 0
(gdb) display s
2: s = 0
(gdb) continue
Continuing.

Breakpoint 1, intm (func=0x400820 <func(float)>, a=0, b=1, n=25)
        at int.cpp:8
8               x += h;
2: s = 5.0922699
1: i = 1
(gdb) print x
$1 = 0.0199999996
(gdb)
```

We can employ the print command to display the values of expressions, as shown in the example. The ptype command prints the type of any variable.

3.6.6 How to step through a program

To execute the next line of the code, we can use the `step` command. In this case, execution of all machine instructions corresponding to one line of a source code continues, and the debugger passes to calling functions. A similar action is performed using the `next` command, but a function call is treated as one line, and the command is executed until exiting this function.

```
(gdb) next
9               s += func(x);
2: s = 5.0922699
1: i = 1
(gdb) next
7          for (int i = 0; i < n; i++) {
2: s = 9.97770882
1: i = 1
(gdb)
```

The commands `nexti`, `stepi` differ from `next`, `step`; only one assembler instruction is executed in them.

3.7 The make utility

The main purpose of the **make utility** is to provide automatic compilation of source codes composed of many files into object files and the following linking into executable files or libraries. On the basis of the information about the time of the last change of a single file, the `make` utility determines and runs the necessary programs.

3.7.1 Usage of the utility

As a rule, a developed program consists of many files. When we change a single file, the corrected version of an executable file can be obtained by recompiling all the files of a project. This is irrational because we have changed only one file. The `make` utility easily resolves this problem.

The `make` utility works with a text fil, which contains all the necessary instructions for working with files. This file is called **makefile**.

The command

```
$ make
```

first searches for the file with name `makefile` in the current directory. If it is not found, then the file `Makefile` is searched. If this file also does not exist, then `make` stops. If for the instruction file some different name is given, e.g. `project`, then `make` is run with the `-f` option:

```
$ make -f project
```

3.7.2 Example

The test program consists of 3 files: `int.cpp`, `int.h`, and `test.cpp`. In the `project` directory, these source files are placed into the `src` directory (see Figure 3.5). The executable file `test` and instruction file `makefile` are also contained in the `project` directory.

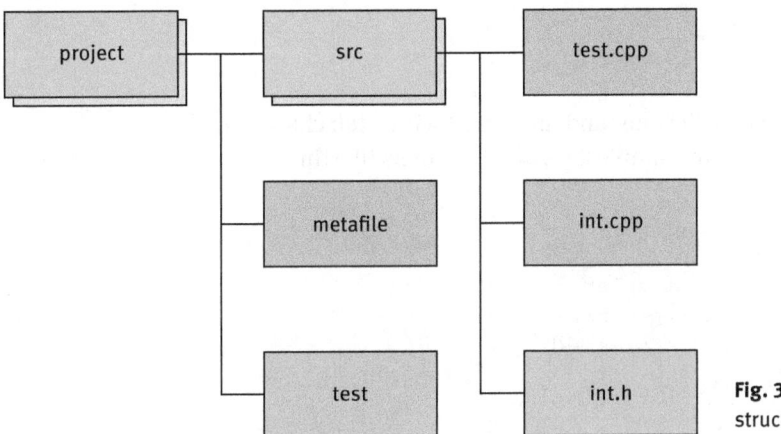

Fig. 3.5. Project structure.

To compile the source files and create an executable file `test` in the `project` directory, we should go to the `project` directory and run the following command:

```
$ g++ -o test src/test.cpp src/int.cpp
```

This single-command approach is acceptable for our project, if extra recompilation costs do not matter. But this is impractical for large projects with many large-size files.

For our project, we can highlight the following intermediate steps of compiling and linking:

```
$ g++ -c -o int.o src/int.cpp
$ g++ -c -o test.o src/test.cpp
$ g++ -o test test.o int.o
```

3.7.3 Simple makefile

Now we create makefile for our project. This file contains **sections** for targets, dependencies, and commands. In makefile, they are arranged as follows: first, the name of the target is defined (usually it is the name of an executable or object file), followed by the : delimiter; secondly, the names of dependencies (files required for this target) are specified. Further, lines are connected with a list of commands, which must be done to achieve this target.

The structure of the file is

```
targets: dependencies
        command
        command
    ...
```

It is assumed that each command must start with a tab character.

For our project, the simplest makefile looks like this:

```
# /project/makefile
test : test.o int.o
    g++ -o test test.o int.o
test.o : src/test.cpp src/int.h
    g++ -c -o test.o src/test.cpp
int.o : src/int.cpp src/int.h
    g++ -c -o int.o src/int.cpp
```

Further, in the project project, the make utility should be run:

```
$ make
g++ -c -o test.o src/test.cpp
g++ -c -o int.o src/int.cpp
g++ -o test test.o int.o
```

Lines that begin with the # character are comments. Restart of make (without changes in the files) does nothing:

```
$ make
make: 'test' is up to date.
```

3.7.4 Phony targets

When we apply the above makefile, the test, test.o, and int.o are created in the project directory. After debugging we need to clean the directory from auxiliary files. This can be done by adding the following rule:

```
clean :
        rm -f *.o
```

to the makefile. No file will appear if we employ the rule clean. To delete files, we can use the command make clean. Similar targets, which are not represented as files, are called phony targets.

This rule does not work if a file named clean exists in the directory with the file makefile. The clean target does not have any dependencies and will always be considered up to date. Here this command for deleting files will never be performed. To resolve this problem, the special **target .PHONY** is used which explicitly declares the phony target

```
.PHONY : clean
```

It seems reasonable to include in makefile the following standard phony targets:
- all – execute all tasks to create a program,
- install – install programs from compiled binary files,
- clean – delete generated binary files etc.

3.7.5 Variables

String **variables** are actively used in applying makefile. Names of variables are given as upper case, as in the following:

```
VAR_NAME=value
```

To get the value of a variable, we need to enclose the name of the variable in parentheses with the $ character at the beginning ($VAR_NAME).

To reduce code from multiple duplication of file names, special automatic variables are employed. For instance, the $@ character is replaced by the current target, the $^ character is replaced by the list of all dependencies with their directories.

The use of variables is illustrated by the following variant of our makefile:

```
# /project/makefile
VPATH = src
SRC_FILES = test.cpp int.cpp
OBJ_FILES = $(patsubst %.cpp, %.o, ${SRC_FILES})
CFLAGS = -c -g
LDFLAGS = -g
.PHONY : clean
test: ${OBJ_FILES}
    g++ ${LDFLAGS} -o test ${OBJ_FILES}
%.o : %.cpp
    g++ ${CFLAGS} -o $@ $^
clean :
    rm -f *.o
```

Here the VPATH variable defines the directory with the project files. Names of object files (the OBJ_FILES variable) are obtained by renaming the files of the project (the patsubst function), the CFLAGS and LDFLAGS variables define compiling and linking with inclusion of debugging information.

Petr N. Vabishchevich

4 The Eclipse IDE in a nutshell

Abstract: The Eclipse[1] Integrated Development Environment (IDE) is widely used to develop cross-platform software written in various programming languages. In this chapter, we study the main features of Eclipse: project management, code editing, compiling, debugging, and linking.

4.1 The Eclipse architecture and GUI

The basic features of Eclipse, its installation, and the various plug-ins are breafly considered below.

4.1.1 Integrated development environments

An **integrated development environment** is a unified system intended for software development. It is designed to quickly create high-quality software (using efficient tools for development and automation) under the best possible conditions (using a convenient graphic user interface (GUI)).

Of course, we can use the command line, write a program in a simple text editor, and debug it using command-line tools. Moreover, some users may like this, but most users prefer to employ IDEs.

Modern IDEs fully resolve various problems in designing software and include a great number of separate components. This results in a certain inconvenience in using development tools and complicates the study and usage of their features.

Therefore, we can begin with some text editor with advanced features which simplify coding. Such editors provide, for example, syntax highlighting and autocompletion, as well as running of the compiler and debugger.

Integrated development environments provide significantly greater convenience and integration in software development. There are a lot of such frameworks for C/C++ programming. The reader can find a long list of these at http://en.wikipedia. org/wiki/Comparison_of_integrated_development_environments.

---

1 http://www.eclipse.org/.

4.1.2 Free cross-platform IDEs

Among existing IDEs, we omit proprietary software (private, commercial). Therefore, our choice is free software (free for use, copying, distribution, and change).

The second requirement is portability of applications, i.e. we need to work simultaneously on Linux and Windows operating systems. From this point of view, we will not consider the contribution of Microsoft to the development of free IDEs such as Microsoft Visual Studio Express[2], which is positioned as an IDE for beginners. We also do not consider e.g. the Anjuta IDE[3], which supports a number of of programming languages (C, C++, Vala, Java, JavaScript, Python), but works only under Linux.

Among cross-platform IDEs, we should mention **Code::Blocks**[4]. This IDE supports GCC compilers as well as GNU GDB. The editor provides syntax highlighting, autocompletion of text, etc. Like other more powerful IDEs, the functionality of Code::Blocks expands by plug-ins.

Geany[5] is positioned as a simple, fast, and compact development environment. It recognizes and highlights the syntax of many programming languages (more than 40) includig, of course, C/C++. Geany supports project management, autocompilation, and execution. This functionality can also be expanded via plug-ins.

A more powerful free IDE, namely, **NetBeans IDE**[6], is written in Java. For this, the Java Virtual Machine (Oracle JDK or an appropriate version of OpenJDK) must be preinstalled in order to employ NetBeans IDE. This IDE is initially focused on the development of applications in Java. Other programming languages (e.g. C/C++, PHP) are supported using plug-ins. However, the user can install a separate IDE installation for the development of C/C++ applications. In particular, NetBeans IDE supports refactoring and profiling tools, as well as visual debugger and a collection of predefined code templates.

The cross-platform library Qt[7] is often used to develop C++ applications with a graphical user interface. The features of this library are most easily applied in the **Qt Creator IDE**. In addition to the standard tools for code writing and debugging applications, Qt Creator is integrated with the visual Qt Designer GUI and the help system Qt Assistant. The Qt Linguist tool, which simplifies the localization and translation of applications into other languages, should also be mentioned.

---

2 http://www.microsoft.com/express.
3 http://www.anjuta.org/.
4 http://www.codeblocks.org/.
5 http://www.geany.org/.
6 http://netbeans.org/.
7 http://qt.nokia.com/.

It seems reasonable to apply **CodeLite**[8] for the development of cross-platform applications with GUI on the basis of the wxWidgets library. This is a free IDE for the C/C++ languages. Project management, autocompletition, refactoring, syntax highlighting, and interactive debugger are also supported in this IDE.

4.1.3 Eclipse

Many developers of cross-platform applications use the free **Eclipse**[9] IDE. In many companies, Eclipse is considered to be the corporate standard of IDE for development in various programming languages.

Eclipse is written in Java and is a platform-independent software. Some users use Eclipse as IDE for the development of applications in Java with the standard tools to work with projects, editor, and debugger. Actually, Eclipse is a general framework for the integration of development tools for working with various types of applications. It is positioned as an extensible platform to develop and support applications throughout their entire life-cycle, i.e. from writing code to supplying updates.

As the platform for various applications, the standard release of Eclipse contains plug-ins which support Java (JDT, Java Development Tools) and allow development of other plug-ins (PDE, Plug-in Development Environment). The user who wants to work with other languages will nees to install special plug-ins. For the most popular languages, istallation packages with built-in plug-ins for the selected language are available.

The extensibility of Eclipse is provided by a set of utilities which are installed as plug-ins. We can select, for example, a preferred system for controling versions as well as various plug-ins for the source code analysis. Thus, using Eclipse we can write codes in any programming language, use any plug-ins, and work on any platform.

The **Eclipse CDT**[10] **C/C++ Development Tools**, the full functionality C/C++ IDE, is built on the basis of the Eclipse platform.

4.1.4 Installation

On the Eclipse download page[11], we can select the **Eclipse IDE** for C/C++ installation package for a specified operating system.

---

8 http://www.codelite.org/.
9 http://www.eclipse.org/.
10 http://www.eclipse.org/cdt/index.php.
11 http://www.eclipse.org/downloads/.

Developers using `Linux` will be interested in the `Eclipse IDE for C/C++ Linux Developers`. It additionally includes plug-ins for debugging: `GCov`, `GProf`, `OProfile` and `Valgrind`.

To run `Eclipse`, we need the `Java Virtual Machine` (`JVM`) to be pre-installed. The version of the installed `Java Runtime Environment` (`JRE`) is checked by the command

```
$ java -version
```

To install `JRE`, the following command

```
$ sudo apt-get install openjdk-7-jre
```

is employed.

The **installation of `Eclipse`** finishes with decompression of the archive into a given directory. The program is invoked by the `eclipse` command (an executable file) in the program's directory.

The first launch of `Eclipse` opens a window for selecting `workspace`, as shown in Figure 4.1. The workspace is a directory for the user's projects. The user can employ the workspace given by default; then this box will no longer appear. To change the workspace, the `File | Switch Workspace` menu command is applied.

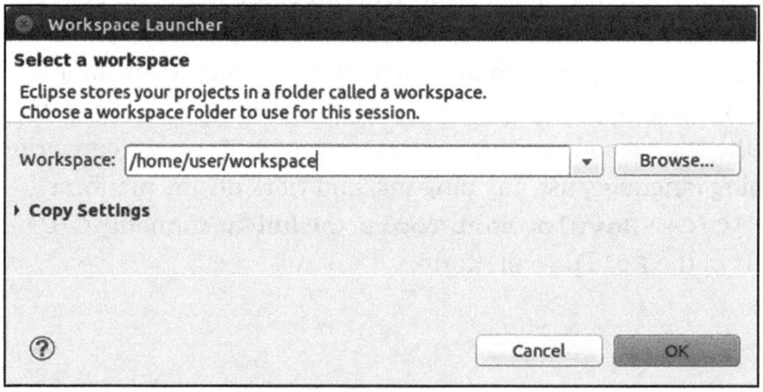

Fig. 4.1. `Eclipse` workspace launcher.

After selecting workspace, the welcome page is displayed (see Figure 4.2). By clicking on one of the buttons, we can go to the page containing
- `Overview` – links to online resources teaching to `Eclipse`;
- `Tutorials` – examples of creating the simplest applications in `C/C++`;
- `Samples` – pre-installed examples;

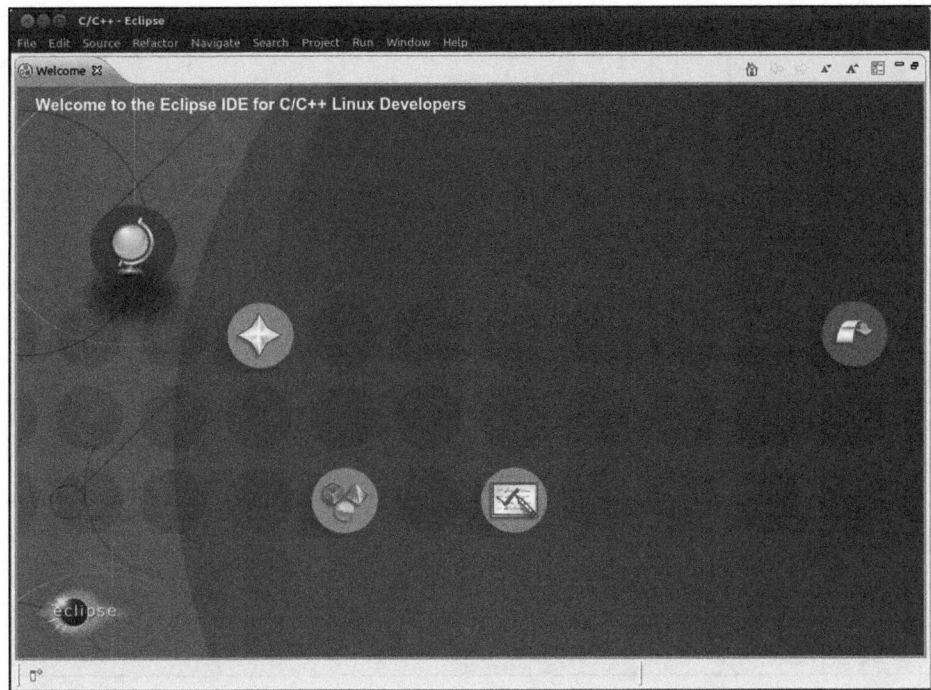

Fig. 4.2. Welcome page.

 item What is new – links to an overview of new features;
– Workbench – the desktop development environment.

To start work, we need to click the Workbench button. The workbench does not contain any projects (see Figure 4.3).

4.1.5 Updating and installing plug-ins

To update Eclipse, we use the Help | Check for Updates menu command.
 To install new plug-ins, the Help | Install New Software menu command (see Figure 4.4) is invoked by selecting a site, from which you want to install the new plug-ins. To add a new site into a list of updates, click the Add button and enter the corresponding address of the site.
 We can address to the unified repository with **Eclipse plug-ins**, i.e. Eclipse Marketplace[12]. Eclipse supports work with this repository and makes it easy to

---

12 http://marketplace.eclipse.org/.

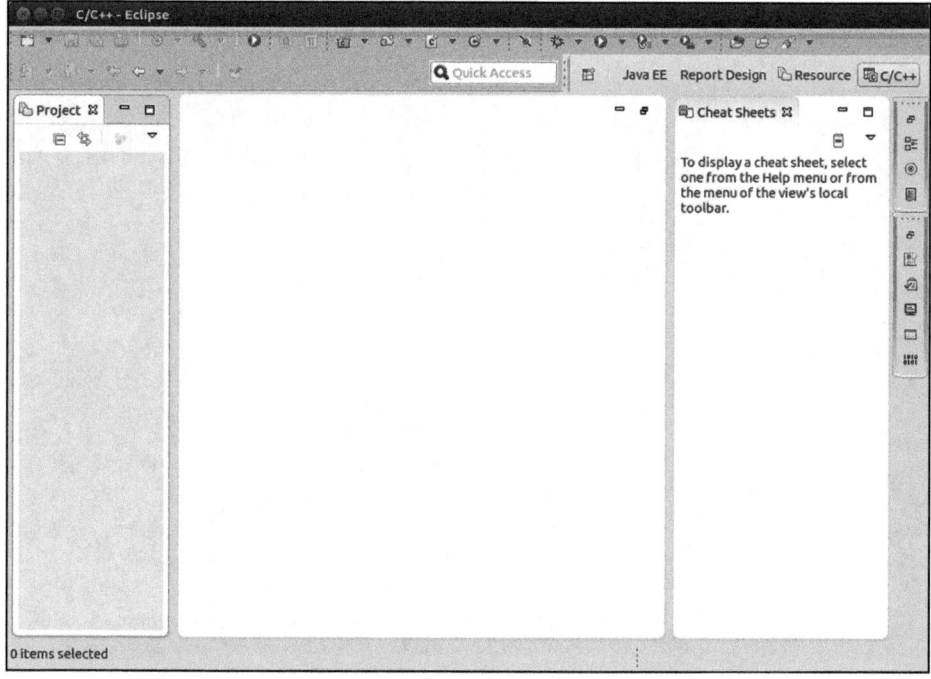

Fig. 4.3. Workspace for the first launch.

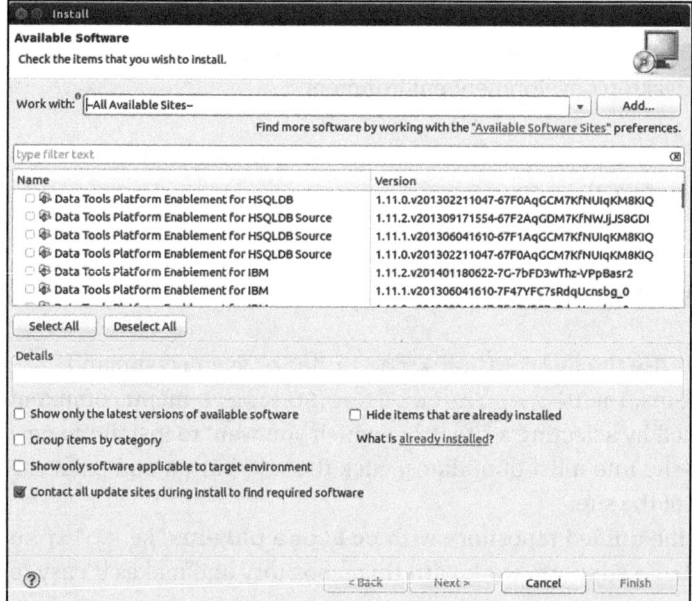

Fig. 4.4. Selection of plug-ins for installation.

Fig. 4.5. Eclipse Marketplace.

search, select, install, and update plug-ins. Using the Help | Eclipse Marketplace menu command, the Eclipse Marketplace dialog is opened, as shown in Figure 4.5.

For instance, it is convenient to represent the results of calculations in the form of graphs, which can be stored as files in a particular graphical format. To visualize such data directly in Eclipse, the QuickImage utility for viewing images is useful. It supports the .gif, .jpg, .jpeg, .png, .bmp, and .ico formats and can be installed from Eclipse Marketplace.

We can get answers to many questions in the advanced Eclipse help system. For example, the help for all integrated plug-ins (see Figure 4.6) is obtained using the Help | Help Contents menu command. In particular, the help on the Eclipse development platform itself is available.

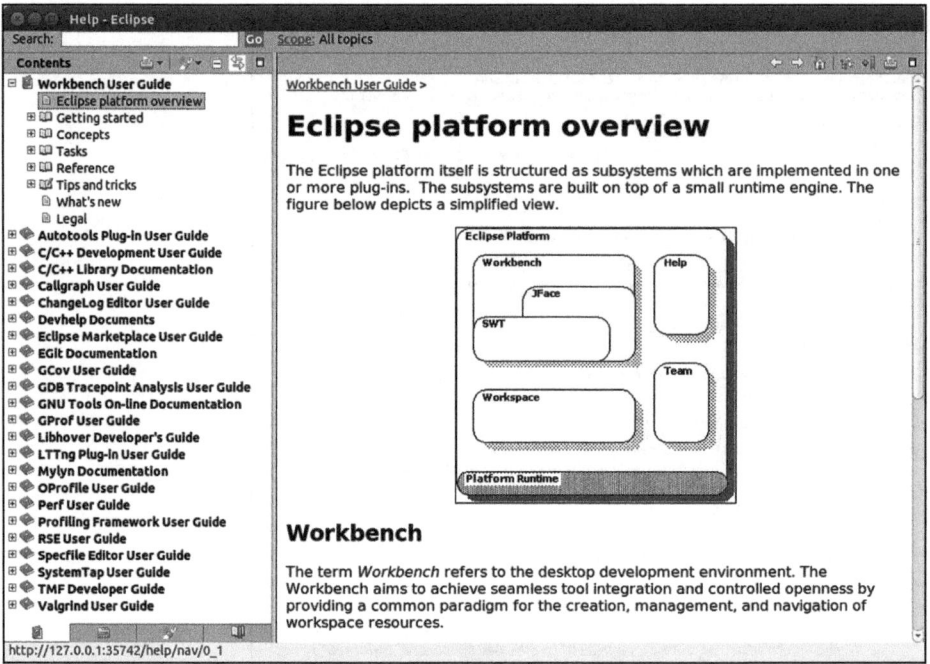

Fig. 4.6. Help for `Eclipse`.

4.2 Working with projects

This section analyzes the creation, compilation, and linking of `C/C++` projects in `Eclipse`.

4.2.1 Available tools

The working tools are the basic `Eclipse`'s modules; they provide the work with projects and help. To resolve special problems, a set of modules is added by activating the corresponding plug-ins.

A defined kit of tools designed to solve specific problems and located in the `Eclipse` Workbench is called **perspective** (layout). Only one perspective can be active at the same time. It is possible to switch perspectives using the `Window | Open Perspective` menu command or the corresponding button in the toolbar. Using perspectives, we can create our own workbench oriented to a certain type of tasks, which takes into account our preferences in both functionality and interface.

Most perspectives consist of the editor area and one or several views. An editor is a tool which allows us to create/open, edit, and save files. In `Eclipse`, different editors can be linked and applied to various types of files. Views supplement the editors and display information about the file we are editing. The Window | Show View menu command is used to open the corresponding view.

4.2.2 Test project

The work with a project in Eclipse is illustrated by solving via the finite difference method the following boundary value problem:

$$\frac{d^2u}{dx^2} = f(x), \quad 0 < x < l,$$

$$u(0) = 0, \quad u(l) = 0.$$

We apply a uniform mesh. To check the accuracy, we consider the exact solution

$$u(x) = \frac{1}{2}(1 - x)x^2, \quad l = 1.$$

To solve a system of equations with a tri-diagonal matrix, the sol function is employed. The source codes, which are contained in the files prog.cpp, prog.h, are given below. These two file are ready to use; we write in the Eclipse editor only the main program.

Listing 4.1.

```
1  // File: prog.h
2  void sol(int n, float *a, float *d, float *c, float *b, float *x);
```

Listing 4.2.

```
1  // File: prog.cpp
2  void sol(int n, float *a, float *d, float *c, float *b, float *x) {
3      int i;
4      float r;
5      for (i = 2; i <= n; i++) {
6          r = a[i - 1] / d[i - 1];
7          d[i] -= r * c[i - 1];
8          b[i] -= r * b[i - 1];
9      }
10     x[n] = b[n] / d[n];
11     for (i = n - 1; i >= 1; i--)
12         x[i] = (b[i] - c[i] * x[i + 1]) / d[i];
13 }
```

We write the `main.cpp` file of the program for solving the boundary value problem by means of `Eclipse`:

Listing 4.3.

```
1   // File: main.cpp
2   #include <iostream>
3   #include <math.h>
4   #include "prog.h"
5   using namespace std;
6   inline float f(float x) {
7       return 1 - 3 * x;
8   }
9   inline float u(float x) {
10      return 0.5 * x * x * (1 - x);
11  }
12  int main() {
13      const int n = 50;
14      const float l = 1.0;
15      float a[n + 1], b[n + 1], c[n + 1], d[n + 1], y[n + 1];
16      int i;
17      float error, h;
18      h = l / n;
19      for (i = 1; i < n; i++) {
20          a[i] = -1.0;
21          d[i] = 2.0;
22          c[i] = -1.0;
23          b[i] = -h * h * f(i * h);
24      }
25      sol(n - 1, a, d, c, b, y);
26      error = 0.0;
27      for (i = 1; i < n; i++)
28          error += (y[i] - u(i * h)) * (y[i] - u(i * h));
29      error = sqrt(error * h);
30      cout << "n = " << n << ", error = " << error;
31      return 0;
32  }
```

Thus, the project consists of three files: `main.cpp`, `prog.cpp`, and `prog.h`.

4.2.3 Creating a project

A project is a folder that contains source codes along with compiling and linking results. Let us now **create a new project** with the name `test`. To do this, we use the `File | New | C++ Project` menu command (the `New` button with a drop-down list

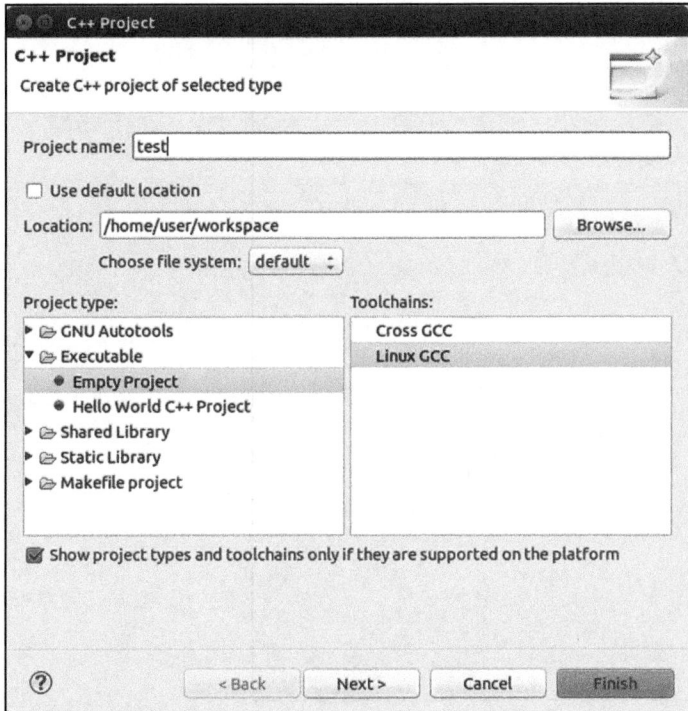

Fig. 4.7. New project.

in the toolbar). In the C++ project wizard, which is shown in Figure 4.7, we set the name and type of our project (in our case, Empty Project without files) and select the Linux GCC toolchain. Next, we use the default settings and click the Finish button. The test project will appear in the Project Explorer view.

Now we create the empty file main.cpp. To do this, in the Project Explorer view, we right-click on the name of our project and select New | File in the context menu (see Figure 4.8).

Next, we need to add the existing files prog.cpp and prog.h to the project. For this we use the Import Wizard that is invoked from the context menu of the project in the Project Explorer; we search for the necessary files in File System (see Figure 4.9). Further, we select the folder and the importing files (see Figure 4.10). The result is shown in Figure 4.11.

The general settings of the Eclipse platform can be edited in the Preferences dialog (the Window | Preferences menu command). The general settings of the development environment are grouped in the General section, settings for C/C++ development are contained in the C/C++ section, and settings for the help system are grouped in the Help section, the Install/Update section contains settings for updating/auto-updating the Eclipse development platform.

Fig. 4.8. New file.

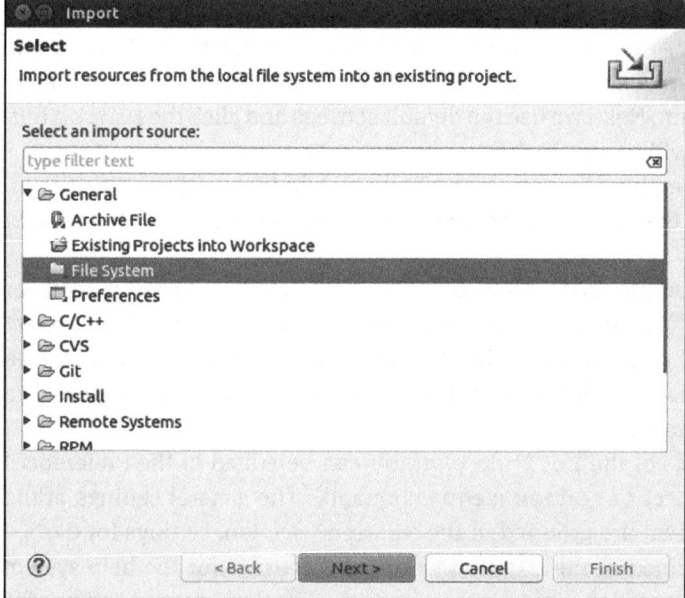

Fig. 4.9. Select files into the project.

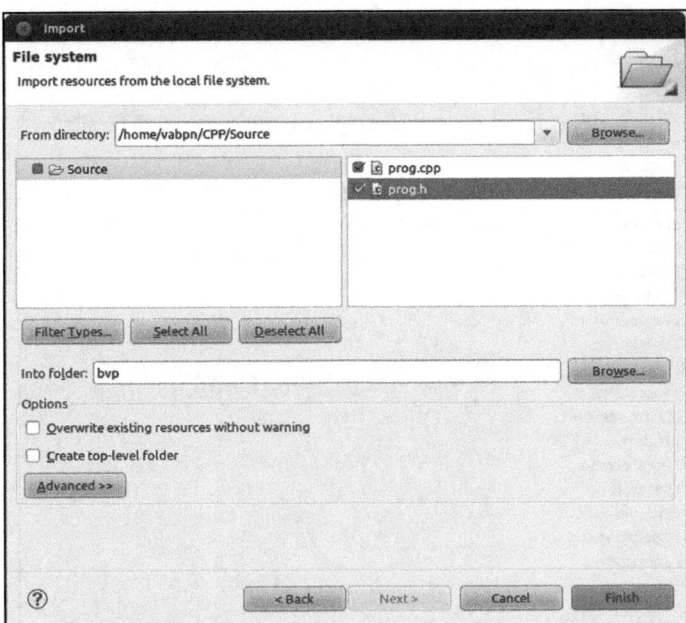

Fig. 4.10. Importing files.

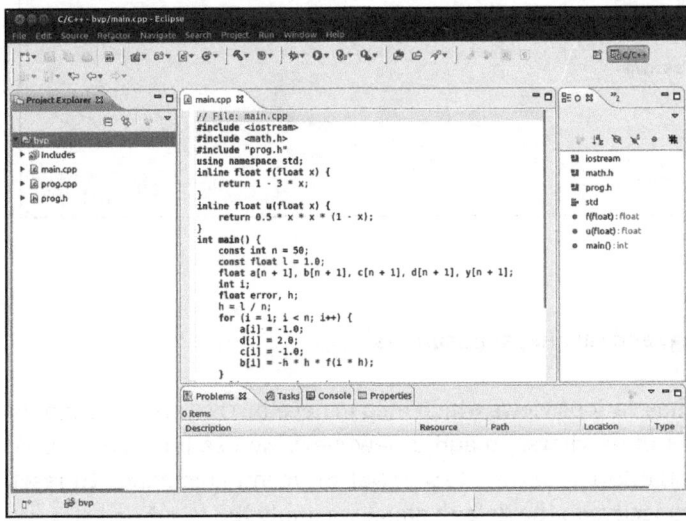

Fig. 4.11. The test project.

Project settings are available in the context menu Properties of the project. In particular, we can obtain the information about compiling and linking commands as shown in Figure 4.12.

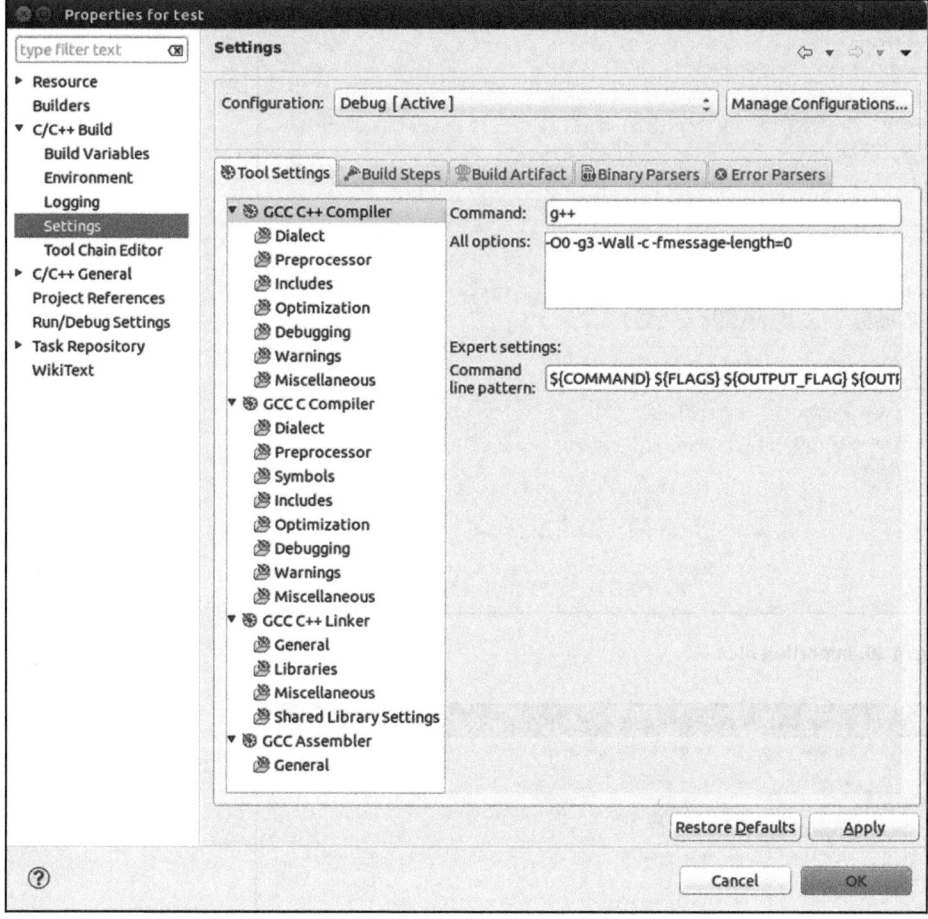

Fig. 4.12. Project parameters.

4.2.4 Compiling, linking, and running applications

The C/C++ perspective has various views which help to develop C/C++ applications. Figure 4.11 demonstrates these views. To add a new view, we use the View Show dialog invoked by the Window | Show View | Other menu command. To reset the current views of perspective to the default state, we apply the Window | Reset Perspective command.

The common views used in the C/C++ perspective are as follows:
- Project Explorer: shows all files in your workspace folder;
- Editor: view to input and edit source files;
- Outline: displays the structure of the current open file;

- `Console`: displays outputs appearing during the linking project and running application;
- `Problems`: shows errors appearing during the work with project.

In `Eclipse` platform, the work of the `C/C++ Development Toolkit (CDT)` is based on including external utilites, which we have discussed above. The automation of the compiling and linking processes is implemented via `make` using the `gcc` or `g++` compilers from the `GCC` collection, whereas `GDB` is applied for debugging.

Building a project is conducted using the `Project | Build Project` menu command (or the `Build Project` context menu of the project) is used for the active configuration (`Debug`, by default). To change the configuration, we use the `Project | Build Configuration | Set Active` menu command. The `Project | Clean` command calls the cleaning dialog to discard the results of the previous project builds.

After linking the project, the configuration folder `Debug` is created in the project folder. This folder contains e.g. object and executable files, as well as the `makefile` file.

An executable file of a project can be invoked with various parameters, e.g. the arguments of the command line. The standard configuration of `C/C++ Application` for an executable file without parameters is generated during the first run of the application using the `Run | Run As` menu command. The generated run configuration is used in subsequent program runs; we can also apply the `Run | Run` menu command or the corresponding button in the toolbar.

In `Eclipse`, parameters required to run an application are specified in the run configuration (the `Run | Run Configuration...` menu command).

4.3 Editing a source code

The editor of `Eclipse` supports standard features of a text editor desined to write programs. Only its basic features that make working with a program's code easier are discussed below.

4.3.1 Customizing the editor

It is possible to customize the **editor** as well as any tool of the `Eclipse` platform. For this purpose, there exists the command `Window | Preferences | C/C++ | Editor`. In particular, we can change the syntax coloring for higher readability (see Figure 4.13).

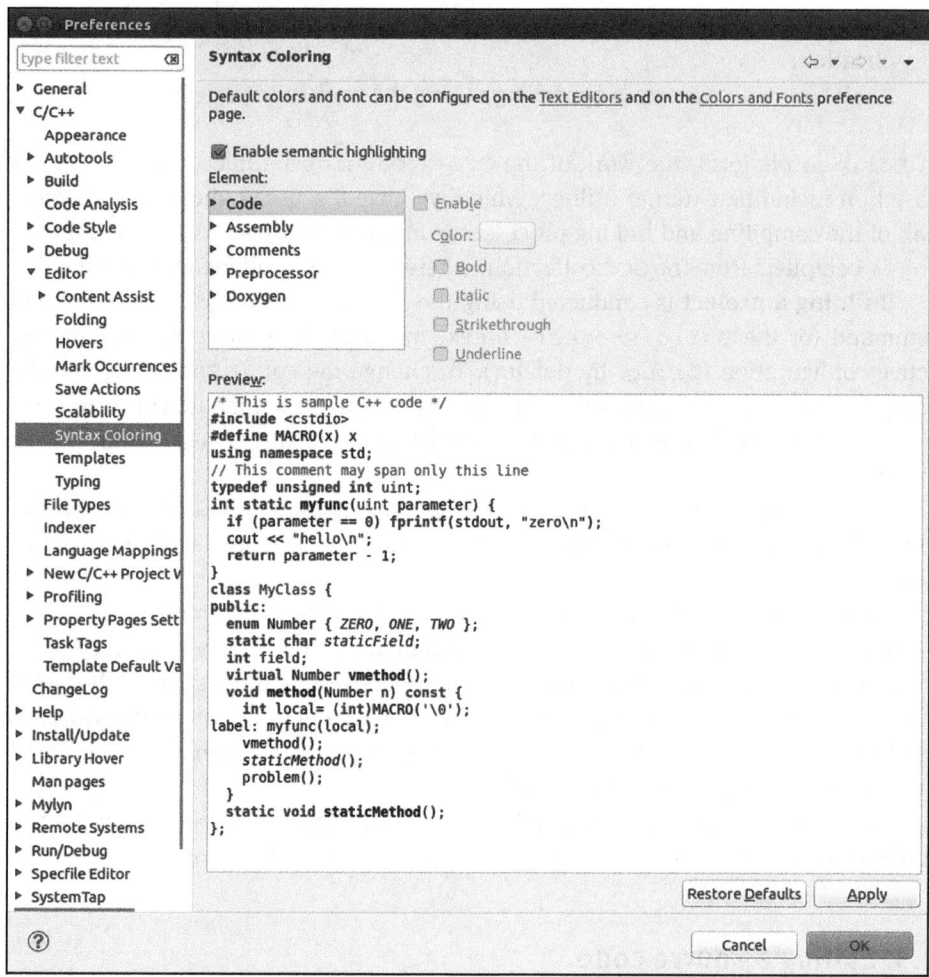

Fig. 4.13. Editor preferences.

We can also customize the keyboard shortcuts using the `General | Keys` section in `Preferences`. For quick access to the list of shortcuts, we click `Ctrl+Shift+L` (see Figure 4.14).

4.3.2 Working with text

We will not discuss in detail the standard commands for working with text, such as selecting, cutting, and copying text. They are available in the `Edit` menu.

Among the useful functions of the `C/C++` code editor, we consider commands for working with comments in the `Source` menu. In particular, to transform a text block

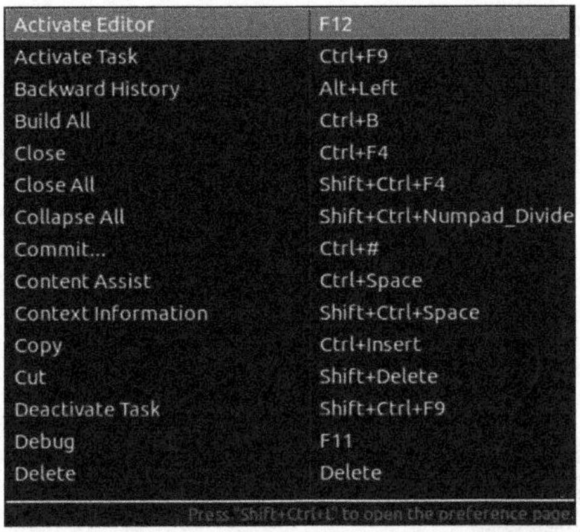

Activate Editor	F12
Activate Task	Ctrl+F9
Backward History	Alt+Left
Build All	Ctrl+B
Close	Ctrl+F4
Close All	Shift+Ctrl+F4
Collapse All	Shift+Ctrl+Numpad_Divide
Commit...	Ctrl+#
Content Assist	Ctrl+Space
Context Information	Shift+Ctrl+Space
Copy	Ctrl+Insert
Cut	Shift+Delete
Deactivate Task	Shift+Ctrl+F9
Debug	F11
Delete	Delete

Press 'Shift+Ctrl+L' to open the preference page.

Fig. 4.14. Shortcuts.

into comments (via `/* ... */`), we can use the `Ctrl+Shift+/` shortcut or the command `Source | Add Block Comment`. In turn, the `Ctrl+Shift+\` or `Source | Remove Block Comment` commands serve to remove this comments.

Regarding the basic features of refactoring (changing the structure of a program without altering its functionality), we need to mention renaming. This allows us to easily rename classes, methods, and variables in all the files of a project. For this purpose, we select an object in the editor view and use the context menu command `Refactor | Rename...` for renaming. The second way is to select the object in the `Project Explorer` and employ the corresponding command from the context menu.

In team development of software, it is necessary to adhere to a single style of programming. This makes the code more readable and understandable for all developers. `Eclipse` has a convenient function for code formatting, i.e. the menu command `Source | Format`. We can choose the formatting style or fine-tune it in the `C/C++ | Code Style` section of the `Preferences`.

The editor also supports code folding. We can fold a block of a code which occupies a large space in the editor view; this makes the code more readable. To fold method bodies, we need to click the minus icon in the left column; to unfold it, we click the plus icon. We can customize code folding in `Windows | Preferences | Editor | Folding`.

4.3.3 Quick insertion

The autocompletion function is realized in the `Content Assist`. To get help on this matter when writing code, we need to click `Ctrl+Space`.

For instance, if we type `erro` and press `Ctrl+Space`, then a list of possible completios appears in our code, as presented in Figure 4.15. Customization of `Content`

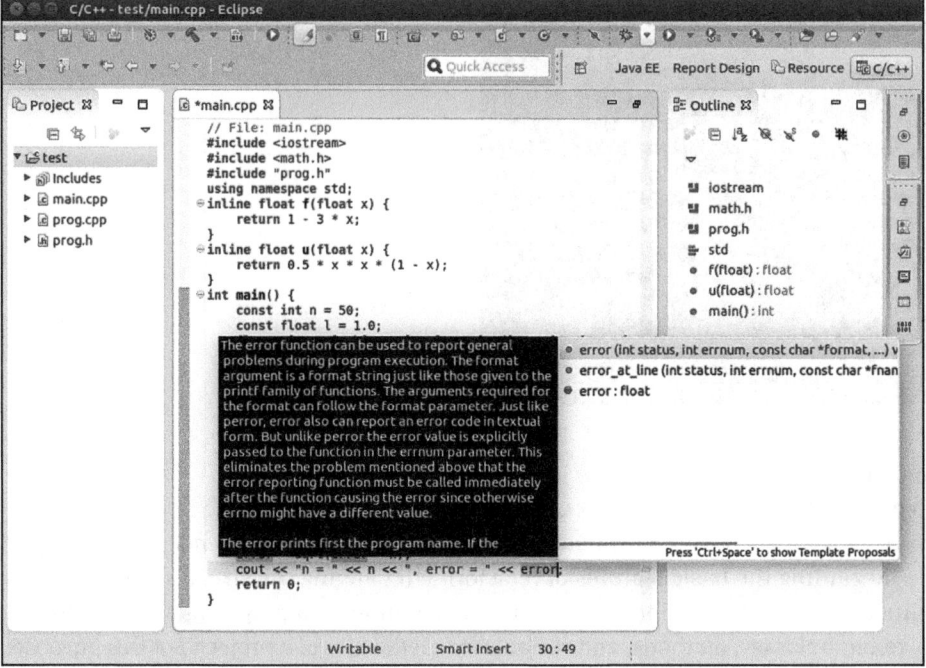

Fig. 4.15. Autocompletion.

`Assist` is conducted in the dialog window activated using the menu command `Windows | Preferences | Editor | Content Assist`.

The editor also supports the feature of quick insert of code sections. For example, to write a loop, we enter `for` and select `Ctrl+Space`, and then select the required template (see Figure 4.16). The list of built-in templates is available through the command `Windows | Preferences | Editor | Templates`. This list can be supplemented with our templates using the `New...` button, which invokes the dialog `New Template`.

4.3.4 Searching for text

`Eclipse` supports searching for text through a specific file, selected files, folders, and the whole project.

The standard `Find/Replace` dialog (see Figure 4.17) is invoked by the command `Edit | Find/Replace` or by pressing `Ctrl+F`. If we choose the `Wrap Search` option, the search is performed cyclically. If it is not chosen, then the search is performed in a part of the file, depending on the cursor position and `Direction` (`Forward` or `Backward`, respectively).

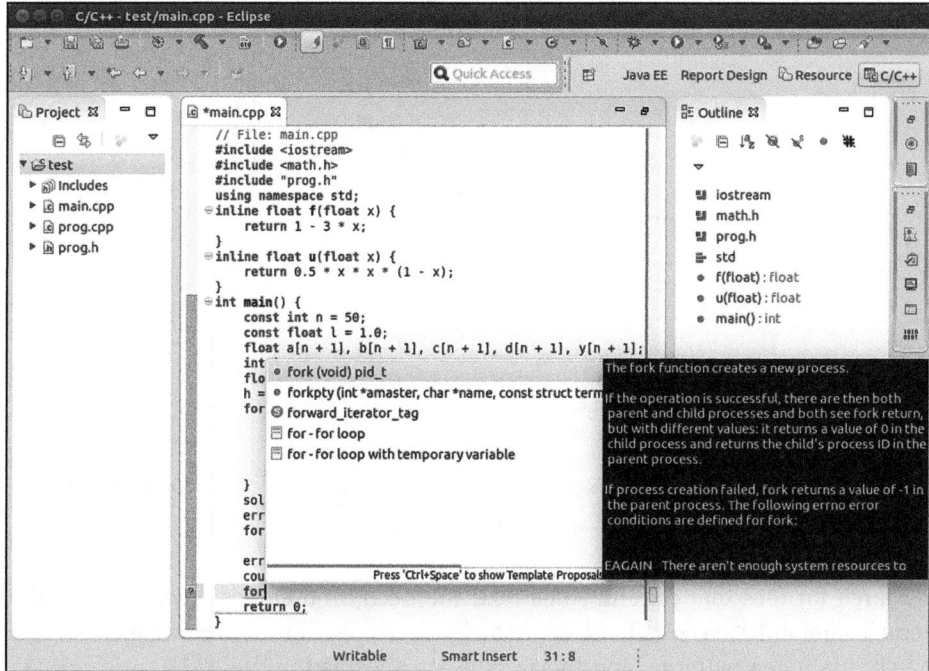

Fig. 4.16. Working with text templates.

Fig. 4.17. Find-and-replace dialog.

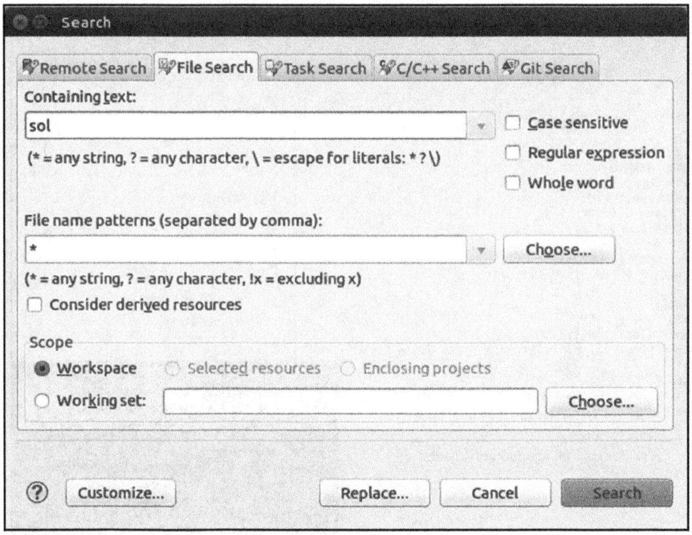

Fig. 4.18. Search for files.

To search for files that contain some text, the `Search` dialog is invoked using the menu command `Search | Search` or by pressing `Ctrl+H`. Parameters of the search, file name patterns, and their scope are specified in the `File search` tab (see Figure 4.18). We can search through the whole workspace or in a set of some selected folders and files (`Working Set`).

To search for a selected text in the editor window, we can apply the command `Search | Text`. The results of the search are collected in the separate `Search` tab (see Figure 4.19), which has convenient buttons to navigate through the results.

4.3.5 Navigation through a code

The `Navigation` menu provides powerfull features for navigating. We mention breafly only some of them.

We can move to a line of a file with a specified number. The corresponding dialog is invoked by the command `Navigation | Go to Line...` (`Ctrl+L`). In some cases, the command `Navigation | Last Edit Location` (`Ctrl+Q`) is useful for moving to the location of a recent edit. To move to the previously opened files (the history of the opened files), commands in the `Navigation | Back` and `Navigation | Forward` menus are used.

To navigate to a declaration of a function, class, or variable, we select the necessary identifier and press `F3`. The same result can be achieved by left-clicking the mouse while pressing `Ctrl`.

In `Eclipse`, various notes in a code are treated as annotations. For example, this may be errors in a code, the result of a search, and so on. Using the navigation icons in

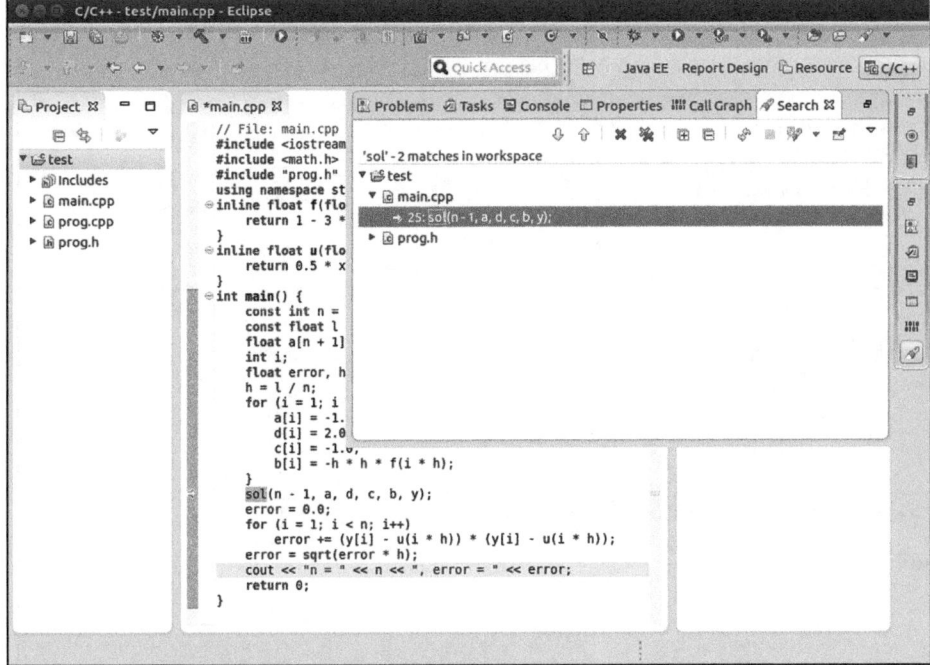

Fig. 4.19. Results of the search.

the toolbar of (`Navigation | Next Annotation` and `Navigation | Previous Annotation`), we can quickly navigate between annotations in the editor window.

4.4 Debugging applications

An introduction to the basic features of `Eclipse` in the detection and localization of errors is presented in the following. Debugging is based on the step-by-step execution of a program with breaks at some lines of the source code. Debugging in `Eclipse` CDT is performed using the `GDB` debugger.

4.4.1 Perspective for debugging

The debugger provides the step-by-step execution of a program and gives an opportunity to set breakpoints, pause, and to resume the program, and observe variables.

The **debugger** of `Eclipse` is organized as the standard set of plug-ins included in `Eclipse`. To open the debug perspective, we use the `Window | Open Perspective | Debug` menu command or the corresponding icons on the perspec-

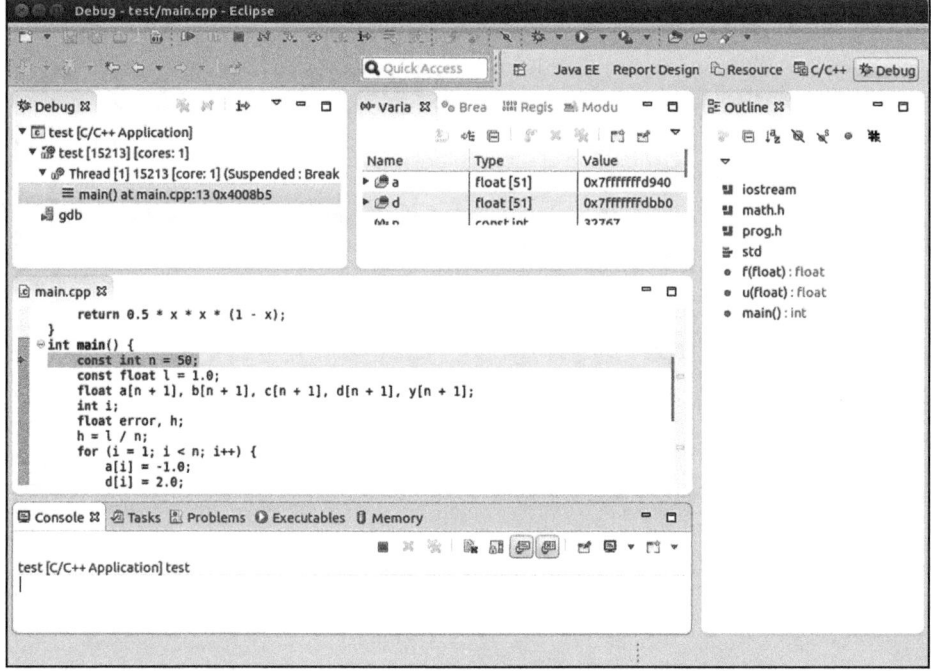

Fig. 4.20. Debug perspective.

tive tab. Figure 4.20 displays the Debug perspective. Switching the mode of working with an application is governed by commands of the Run | Run or Run | Debug menues.

Editing a debug configuration is conducted using the Debug Configurations dialog, which is invoked by the command Run | Debug Configurations... (see Figure 4.21).

Debugging a program and searching for errors, in many respects, are an art, but this does not mean that the programmer does not adhere in his work to some general rules and principles. Here we show only some basic tips and tricks for debugging.

4.4.2 Breakpoints

Debugging an application is based on, in particular, the intentional interruption using breakpoints. Using the debugger, we can examine a state of the program at the time moment of interruption. After that, the program can be completed or resumed from the point where it was terminated.

To create a breakpoint, we should select a line of the code and right-click on the left margin of the editor near the necessary line. After that, we choose Toggle

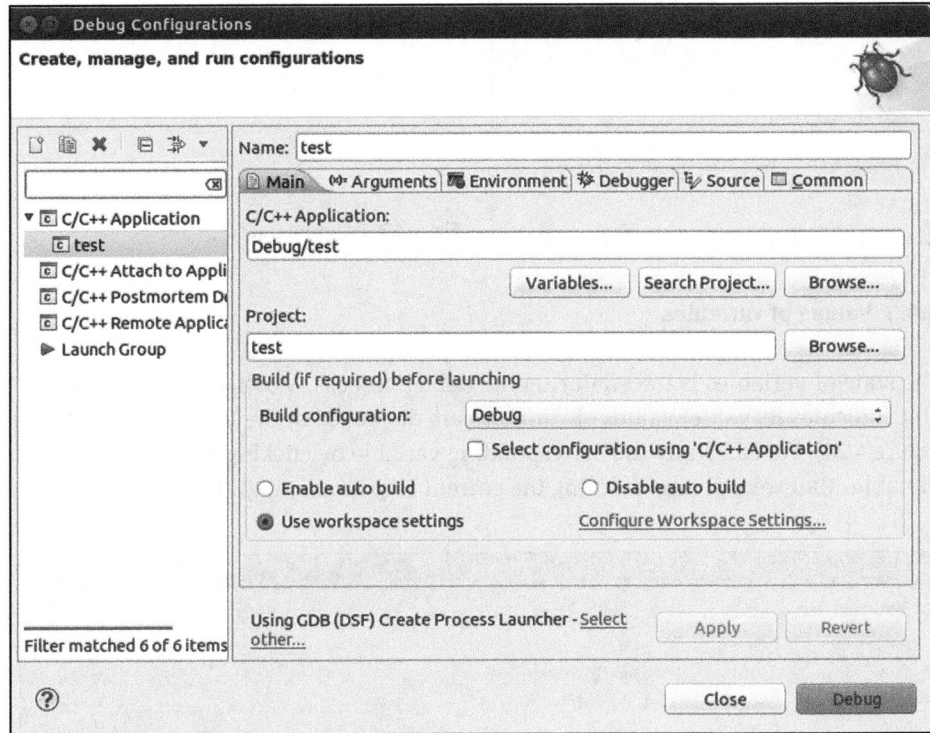

Fig. 4.21. Debug configuration.

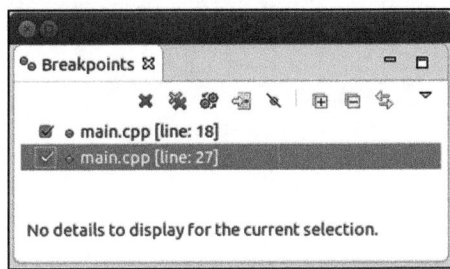

Fig. 4.22. Breakpoints.

Breakpoint in the context menu or use the Run | Toggle Breakpoint menu command. Managing a list of breakpoints for a project is performed on the Breakpoints tab of the Debug perspective (see Figure 4.22).

The commands Run | Step Into, Run | Step Over, and Run | Step Return are used to step through a code. The line of the code that will be executed next is selected for stepping. We can use the corresponding icons in the Debug tab (see Figure 4.23). To move to the next breakpoint, the Run | Resume command is applied.

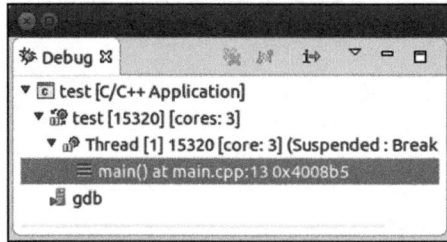

Fig. 4.23. Program trace.

4.4.3 Values of variables

The state of variables is traced during the debug process. Moving cursor over a variable indicates its value. Values of variables are displayed in the `Variables` tab (see Figure 4.24). We can study the corresponding variable by clicking its name in the list. Variables that were changed during the current step are highlighted.

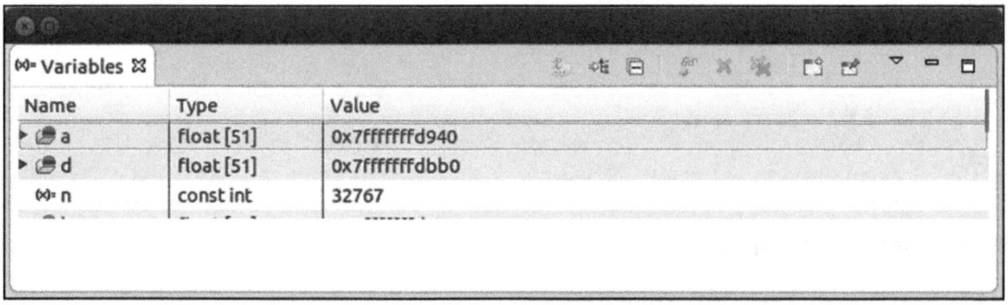

Name	Type	Value
▶ a	float [51]	0x7fffffffd940
▶ d	float [51]	0x7fffffffdbb0
(x)= n	const int	32767

Fig. 4.24. Variables.

Nadezhda M. Afanasyeva, Victor S. Borisov, Maria V. Vasilieva,
Aleksandr V. Grigoriev, Petr E. Zakharov, Petr A. Popov, and
Ivan K. Sirditov

5 The GSL scientific library

Abstract: The **GNU Scientific Library (GSL)**[1] is a collection of numerical tools for solving basic problems of computational mathematics. The library is written in the C programming language and distributed under the GNU General Public License[2].

5.1 Preliminaries

The GSL library provides a large number of functions (more then 1000) for the numerical solution of systems of linear algebraic equations and ordinary differential equations (ODEs), as well as functions for working in such areas of computational mathematics as combinations, statistics, optimization etc.

To **install GSL,** we need the C compiler of the ANSI C standard. The library can be installed from the source code (using the make tool[3]) or from the Ubuntu repository using the command:

```
$ sudo apt-get install libgsl0-dev
```

To run a program that employs the GSL possibilities, we need to compile a source code

```
$ gcc -I/usr/local/include -c example.c
```

and then link the program

```
$ gcc -L/usr/local/lib example.o -lgsl -lgslcblas
```

The result is the executable file.

---

1 http://www.gnu.org/software/gsl/.
2 http://www.gnu.org/licenses/gpl.html.
3 http://www.gnu.org/software/make/.

5.2 Functions and constants

The GSL library provides the standard and special mathematical functions, statistical functions, physical constants, and generators of random numbers.

5.2.1 Mathematical and special functions

The GSL library provides **standard mathematical functions** as an alternative to system libraries; they can be used when the system functions are not available.

A short list of mathematical constants is given in Table 5.1. Table 5.2 presents the standard functions implemented in the library. To work with these functions and constants, it is necessary to include the header file gsl_math.h.

Table 5.1. Mathematical constants.

Name	Description
M_E	The base of exponentials, e
M_LOG2E	The base-2 logarithm of e
M_LOG10E	The base-10 logarithm of e
M_SQRT2	The square root of two
M_SQRT3	The square root of three
M_PI	The constant π
M_LN10	The natural logarithm of 10
M_LN2	The natural logarithm of 2
M_LNPI	The natural logarithm of π
M_EULER	Euler's constant

Table 5.2. Standard functions.

Name	Description
double gsl_log1p (const double x)	The function log(1 + x)
double gsl_expm1 (const double x)	The function exp(x) - 1
double gsl_acosh (const double x)	The function arccosh(x)
double gsl_asinh (const double x)	The function arcsinh(x)
double gsl_atanh (const double x)	The function arctanh(x)

In the GSL library, special functions are presented in two forms: a natural form and error-handling form. The natural form returns only the value of a function, and therefore, such a form can be used in mathematical expressions. A function in the error-handling form returns an error code in addition to the value. A structure of returned value is declared in the header file gsl_sf_result.h and defined as follows:

```
typedef struct {
    double val;
    double err;
} gsl_sf_result;
```

Here val is the value of the function and err is an estimate of the absolute error. Functions in the error-handling form have the additional suffix _e.

Now we present an example of how to use a function in the form of error handling functions to calculate the Bessel function J0(x).

Listing 5.1.

```
1   #include <stdio.h>
2   #include <gsl/gsl_errno.h>
3   #include <gsl/gsl_sf_bessel.h>
4   int main(void) {
5       double x = 1.0;
6       gsl_sf_result result;
7       int status = gsl_sf_bessel_J0_e(x, &result);
8       printf("status  = %s\n", gsl_strerror(status));
9       printf("J0(1.0) = %.18f\n", result.val);
10      printf("error   = %.18f\n", result.err);
11      return status;
12  }
```

The result of the program is as follows:

```
status  = success
J0(1.0) = 0.765197686557966494
error   = 0.000000000000000673
```

In this program, we employ the function gsl_sf_bessel_J0_e, which calculates the Bessel function with a given x. Other special functions are also available in the library.

5.2.2 Physical constants

The library contains a collection of macros to operate with **physical constants**, such as the speed of light and the gravitational constant. The values are presented in different unit systems, including the MKSA (meters, kilograms, seconds, amperes) and CGSM (centimeters, grams, seconds, gauss) systems.

The header file `gsl_const_mksa.h` contains constants in the MKSA system, constants in the CGSM system are defined in the `gsl_const_cgsm.h`, and dimensionless constants are governed by `gsl_const_num.h`.

The use of physical constants is shown in the following example: we want to calculate the light-travel time from Earth to Mars.

Listing 5.2.

```
1  #include <stdio.h>
2  #include <gsl/gsl_const_mksa.h>
3  int main(void) {
4      double c = GSL_CONST_MKSA_SPEED_OF_LIGHT;
5      double au = GSL_CONST_MKSA_ASTRONOMICAL_UNIT;
6      double minutes = GSL_CONST_MKSA_MINUTE;
7      double r_earth = 1.00 * au;
8      double r_mars = 1.52 * au;
9      double t_min, t_max;
10     t_min = (r_mars - r_earth) / c;
11     t_max = (r_mars + r_earth) / c;
12     printf("minimum = %.1f minutes\n", t_min / minutes);
13     printf("maximum = %.1f minutes\n", t_max / minutes);
14     return 0;
15  }
```

The output is

```
minimum = 4.3 minutes
maximum = 21.0 minutes
```

Here the speed of light in a vacuum is set using GSL_CONST_MKSA_SPEED_OF_LIGHT. The constant GSL_CONST_MKSA_ASTRONOMICAL_UNIT is applied to indicate the length of one astronomical unit, which is approximately equal to the average distance between the Earth and the Sun. The constant GSL_CONST_MKSA_MINUTE is employed to calculate time in minutes.

In addition to the above constants, GSL has other constants, which are presented in Table 5.3.

5.2.3 Random number generators

The GSL library is supplied by a collection of **random number generators**. These generators are accessible through a uniform interface. Environment variables permit quickly switching between different generators without the need to recompile a pro-

Table 5.3. Physical constants.

Name	Description
GSL_CONST_MKSA_PLANCKS_CONSTANT_H	Planck's constant
GSL_CONST_NUM_AVOGADRO	Avogadro's number
GSL_CONST_MKSA_FARADAY	The molar charge of 1 Faraday
GSL_CONST_MKSA_GRAVITATIONAL_CONSTANT	Gravitational constant
GSL_CONST_MKSA_MICRON	The length of 1 micron
GSL_CONST_MKSA_TON	The mass of 1 ton
GSL_CONST_MKSA_CALORIE	The energy of 1 calorie
GSL_CONST_MKSA_STD_ATMOSPHERE	The pressure of 1 standard atmosphere
GSL_CONST_MKSA_NEWTON	The force of 1 Newton
GSL_CONST_MKSA_GAUSS	The magnetic field of 1 Gauss

gram. Functions for working with random numbers are declared in the header file `gsl_rng.h`.

The following example demonstrates some features of the library in using generators.

Listing 5.3.

```
1  #include <stdio.h>
2  #include <gsl/gsl_rng.h>
3  int main(void) {
4      const gsl_rng_type * T;
5      gsl_rng * r;
6      int i, n = 10;
7      gsl_rng_env_setup();
8      T = gsl_rng_default;
9      r = gsl_rng_alloc(T);
10     for (i = 0; i < n; i++) {
11         double u = gsl_rng_uniform(r);
12         printf("%.5f\n", u);
13     }
14     gsl_rng_free(r);
15     return 0;
16  }
```

The function gsl_rng_type * gsl_rng_env_setup reads the environment variables GSL_RNG_TYPE, GSL_RNG_SEED and writes the obtained values in the global variables gsl_rng_default and gsl_rng_default_seed, where gsl_rng_default is a pointer to a random number generator. The function gsl_rng_alloc initializes the random number generator. Further, in the loop we obtain a sequence of random numbers using the function gsl_rng_uniform, which returns a number uniformly distributed in the range from 0 to 1. Finally, the function gsl_rng_free frees the memory allocated by the random number generator.

The output of the program is as follows:

```
0.99974
0.16291
0.28262
0.94720
0.23166
0.48497
0.95748
0.74431
0.54004
0.73995
```

The numbers depend on a key used by the generator. The key can be changed using the environment variable GSL_RNG_SEED. The generator can also be changed using the environment variable GSL_RNG_TYPE.

Other generators are also available in GSL (see Table 5.4).

Table 5.4. Types of random number generator.

Name	Description
gsl_rng_mt19937	Mersenne Twister (MT19937)
gsl_rng_ranlxs0	The generator ranlxs0
gsl_rng_cmrg	Multiple recursive generator by L'Ecuyer
gsl_rng_taus	Tausworthe generator

These generators are applied to model processes in order to pass statistical tests.

5.2.4 Statistical functions

The library provides **basic statistical functions** to calculate the mean, variance and standard deviation. More advanced functions allow evaluation of absolute deviations, skewness, kurtosis and the median along with arbitrary percentiles.

The functions are available for both integer and floating-point types. For the floating-point data type, we need to use functions with the prefix gsl_stats declared in the header file gsl_statistics_double.h. For the integer data type, we employ a function with the prefix gsl_stats_int declared in the header file gsl_statistics_int.h.

Let us present an example demonstrating the use of the statistical functions from GSL.

Listing 5.4.

```
1  #include <stdio.h>
2  #include <gsl/gsl_statistics.h>
3  int main(void) {
4      double data[5] = {17.2, 18.1, 16.5, 18.3, 12.6};
5      double mean, variance, largest, smallest;
6      mean = gsl_stats_mean(data, 1, 5);
7      variance = gsl_stats_variance(data, 1, 5);
8      largest = gsl_stats_max(data, 1, 5);
9      smallest = gsl_stats_min(data, 1, 5);
10     printf("The dataset is %g, %g, %g, %g, %g\n",
11             data[0], data[1], data[2], data[3], data[4]);
12     printf("The sample mean is %g\n", mean);
13     printf("The estimated variance is %g\n", variance);
14     printf("The largest value is %g\n", largest);
15     printf("The smallest value is %g\n", smallest);
16     return 0;
17  }
```

The results of the program are as follows:

```
The dataset is 17.2, 18.1, 16.5, 18.3, 12.6 The sample mean is 16.54
The estimated variance is 5.373
The largest value is 18.3
The smallest value is 12.6
```

In this example, we apply the function gsl_stats_mean to calculate the arithmetic mean; the function gsl_stats_variance is used to evaluate the variance of random data, whereas the functions gsl_stats_max and gsl_stats_min are employed to obtain maximum and minimum values. Table 5.5 shows the list of statistical functions.

Table 5.5. Statistical functions.

Name	Description
gsl_stats_sd	Standard deviation
gsl_stats_absdev	Absolute deviation from the mean value
gsl_stats_skew	The skewness
gsl_stats_kurtosis	The kurtosis
gsl_stats_covariance	The covariance
gsl_stats_correlation	The Pearson correlation coefficient between elements of an array

5.3 Combinatorics

The GSL library has functions to solve problems in combinatorial analysis. Here we consider typical problems for permutation, combination, and enumeration.

5.3.1 Permutation

In combinatorics, a permutation p is an ordered set of n integers in the range 0 to $n - 1$, where each value p_i occurs only once. For example, the integers $(0, 1, 3, 2)$ represents a **permutation**, where the last two numbers have exchanged positions.

Let us solve the following problem: we want to display all the permutations of the string "republic".

Listing 5.5.

```
1   #include <stdio.h>
2   #include <string.h>
3   #include <gsl/gsl_permutation.h>
4   int main(void) {
5       char s[] = "republic"; // Initial string
6       int n = strlen(s);
7       int i;
8       // Allocate memory and initialize a permutation
9       gsl_permutation* p = gsl_permutation_calloc(n);
10      do {
11          // Print rearranged string by letters
12          for (i = 0; i < n; i++) {
13              printf("%c", s[gsl_permutation_get(p, i)]);
14          }
15          printf("\n");
16          // Construct the next permutation
17      } while (gsl_permutation_next(p) == GSL_SUCCESS);
18      gsl_permutation_free(p); // Free memory
19      return 0;
20  }
```

The program creates a permutation with the size equal to the number of letters in the string. We display the letters in the i-th position of the current permutation. The library function calculates the next permutation and iterations continues until all possible permutations are complete.

Functions for working with permutations use the structure gsl_permutation. This structure is declared as follows:

```
typedef struct {
    size_t size;
    size_t *data;
} gsl_permutation;
```

The pointer `data` refers to an array with `size` integers. Before working with a permutation and, in general, with all structures in GSL, we need to correctly allocate memory calling one of the following functions:

```
gsl_permutation *gsl_permutation_alloc(size_t n);
gsl_permutation *gsl_permutation_calloc(size_t n);
```

Here the parameter `n` is the number of integers in a permutation. The difference between these two functions is that the pointer returned by `gsl_permutation_alloc` must be initialized using the function `gsl_permutation_init`, while `gsl_permutation_calloc` returns the same pointer, but the permutation is initialized. When we finish working with a permutation, we must free the allocated memory using the function

```
void gsl_permutation_free(gsl_permutation *p);
```

To access the `i`-th element of the current permutation, we employ the function

```
size_t gsl_permutation_get(const gsl_permutation *p, size_t i);
```

The value returned by this function can be used as the index in the char array `s`. The next permutation is generated using the function

```
int gsl_permutation_next(gsl_permutation *p);
```

The function rearranges elements of the structure `p` in lexicographic order and returns `GSL_SUCCESS`, if the permutation has been created, or `GSL_FAILURE`, if no further permutations are available.

Thus, the program displays the following result:

```
republic
republci
republic
repubicl
repubcli
...
```

5.3.2 Combinations

A **combination** $\binom{n}{k}$ is a subset of k elements chosen from a given set containing n different elements. The order of elements is not considered. The GSL library contains functions for computing all the possible subsets in a combination.

Let us consider the classic knapsack problem. We have a set of items, each with two parameters: a mass and value. Also, we have a knapsack with a certain capacity. The aim is to collect the knapsack with the maximum total value of the items under the restriction that the total weight is less than or equal to the given limit for the knapsack.

There are different solutions of this problem. The most obvious method is the brute force approach (searching for all possible combinations). Thus, we must create a list of n items and construct combinations $\binom{n}{k}$ for $k \in [1, n]$. For each combination of items, we calculate the total weight and value as well as choose a collection with the maximum value.

The program begins with the inclusion of the GSL library, definitions of constants and structures:

Listing 5.6.

```
1  #include <stdio.h>
2  #include <time.h>
3  #include <gsl/gsl_combination.h>
4  #define NITEM       5
5  #define SACK_WEIGHT 10
6  typedef struct {
7      int weight;
8      int value;
9  } ITEM;
```

The header file gsl/gsl_combination.h contains functions to operate with combinations. The constants NITEM and SACK_WEIGHT denote the number of items and the weight limit of the knapsack, respectively. The structure ITEM is an item with a weight and a value.

The initialization of the array with items is performed using the pseudo-random number generator from the standard library:

Listing 5.7.

```
1  srand(time(NULL));
2  ITEM items[NITEM];
3  int i,k,j;
4  for(i=0; i<NITEM; i++) {
5      items[i].weight = rand() % 10 + 1;
6      items[i].value = rand() % 10 + 1;
7      printf("item %d has weight %d and value %d\n",
8              i, items[i].weight, items[i].value);
9  }
```

The expression `rand() % 10 + 1` represents the generation of a number from $[1, 10]$.

The GSL library uses the structure `gsl_combination` for working with combinations:

```
typedef struct {
    size_t n;
    size_t k;
    size_t *data;
} gsl_combination;
```

The initialization of a combination is performed by means of the function

```
gsl_combination *gsl_combination_calloc(size_t n, size_t k);
```

It returns a pointer to the structure of the combination $\binom{n}{k}$, where the lexicographically first combination is initialized. In the example with the knapsack, to examine all the possible combinations of NITEM with k items, we need to create a pointer

```
gsl_combination *c = gsl_combination_calloc(NITEM, k);
```

To get the `i`-th element of the subset, the following function is used:

```
size_t gsl_combination_get(const gsl_combination *c, const size_t i);
```

To evaluate the `weight` and `val` of items of the current combination c, we employ the following loop:

Listing 5.8.

```
1  int val = 0;
2  int weight = 0;
3  for(j=0; j<k; j++) {
4      val += items[gsl_combination_get(c, j)].value;
5      weight += items[gsl_combination_get(c, j)].weight;
6  }
```

If items from the current subset have the maximum value and its total value is not larger than the knapsack weight, then the set c is stored as max_c. At the same time, we cannot just copy the pointer, since subsets in the combination c are changing. For this, we can use the function

```
int gsl_combination_memcpy(gsl_combination *dest,
                           const gsl_combination *src);
```

Combinations must have the same size; therefore, we must allocate memory before copying the combination:

Listing 5.9.

```
1  if (val > max_val && weight <= SACK_WEIGHT){
2      max_val = val;
3      if (max_c) {
4          gsl_combination_free(max_c);
5      }
6      max_c = gsl_combination_alloc(NITEM, k);
7      gsl_combination_memcpy(max_c, c);
8  }
```

Here gsl_combination_free frees the memory allocated by functions gsl_combination_calloc or gsl_combination_alloc. The latter two functions work similarly to the permutation functions. The memory allocation occur each time, since the number k of combined items can be changed, whereas the size of copying combination must be equal.

The step to the next combination ($\frac{n}{k}$) is performed using the function

```
int gsl_combination_next(gsl_combination *c);
```

Here the returned value is equal to GSL_SUCCESS or GSL_FAILURE, similarly to the permutation. Therefore, the main loop appears like this:

Listing 5.10.

```
1 for (k = 1; k <= NITEM; k++) {
2     gsl_combination *c = gsl_combination_calloc(NITEM, k);
3     do {
4         ...
5     } while(gsl_combination_next(c) == GSL_SUCCESS);
6     gsl_combination_free(c);
7 }
```

Printing the combination with the maximum value is conducted with the function

```
int gsl_combination_fprintf(FILE *stream, const gsl_combination *c,
const char *format);
```

The parameter `format` must contain an integer modifier, e.g. `%d`, which is replaced by the `i`-th element of the combination. All elements are written to the `stream`. In our example, it is the standard output stream `stdout`:

Listing 5.11.

```
1 if (max_val > 0){
2     printf("sack items: ");
3     gsl_combination_fprintf(stdout, max_c, "%d ");
4     printf("\n");
5 }
```

Here is the complete code:

Listing 5.12.

```
1  #include <stdio.h>
2  #include <time.h>
3  #include <gsl/gsl_combination.h>
4  #define NITEM 5
5  #define SACK_WEIGHT 10
6  // Define the structure
7  typedef struct {
8      int weight;
9      int value;
10 } ITEM;
11 int main(void) {
12     srand(time(NULL));
13     ITEM items[NITEM];
14     int i,k,j;
```

```
15      // Generate input data
16      for(i=0; i<NITEM; i++){
17          items[i].weight = rand() % 10 + 1;
18          items[i].value = rand() % 10 + 1;
19          printf("item %d has weight %d and value %d\n",
20                  i, items[i].weight, items[i].value);
21      }
22      gsl_combination *max_c = NULL;
23      int max_val = 0;
24      for(k=1; k<=NITEM; k++) {
25          gsl_combination *c = gsl_combination_calloc(NITEM, k);
26          //The brute force
27          do {
28              // Calculate value
29              int val = 0;
30              int weight = 0;
31              for(j=0; j<k; j++){
32                  val += items[gsl_combination_get(c, j)].value;
33                  weight += items[gsl_combination_get(c, j)].weight;
34              }
35              // Save the maximum value
36              if (val > max_val && weight <= SACK_WEIGHT){
37                  max_val = val;
38                  if (max_c) {
39                      gsl_combination_free(max_c);
40                  }
41                  max_c = gsl_combination_alloc(NITEM, k);
42                  gsl_combination_memcpy(max_c, c);
43              }
44          } while(gsl_combination_next(c) == GSL_SUCCESS);
45          gsl_combination_free(c);
46      }
47      // Print the solution
48      if (max_val > 0){
49          printf("sack items: ");
50          gsl_combination_fprintf(stdout, max_c, "%d ");
51          printf("\n");
52      }
53      return 0;
54  }
```

The output is as follows:

```
item 0 has weight 7 and value 2
item 1 has weight 4 and value 3
item 2 has weight 1 and value 10
item 3 has weight 8 and value 8
item 4 has weight 3 and value 6
sack items: 1 2 4
```

5.3.3 Multisets

A **multiset** is a subset of k elements chosen from a given set containing n various elements, which may appear more than once.

Let us consider the following problem. We need to find all the possible versions of passwords with 3 digits from 0 to 8 and write obtained passwords in the lexicographical order.

Listing 5.13.

```
1  #include <stdio.h>
2  #include <time.h>
3  #include <gsl/gsl_multiset.h>
4  int main(void) {
5      gsl_multiset *m = gsl_multiset_alloc(9, 3);
6      do {
7          gsl_multiset_fprintf(stdout, m, "%d ");
8          printf("\n");
9      } while(gsl_multiset_next(m)==GSL_SUCCESS);
10     gsl_multiset_free(m);
11     return 0;
12 }
```

The program output is as follows:

```
0 0 0
0 0 1
0 0 2
...
7 7 8
7 8 8
8 8 8
```

The program begins with the inclusion of the header file `gsl_multiset.h` and the initialization of the multiset using the function `gsl_multiset_alloc`, where the lexicographically first multiset is initialized. Similarly to permutations and combinations, the step to the next multiset is performed by the function `gsl_multiset_next`. When we finish working with a multiset, we must free the allocated memory using the function `gsl_multiset_free`.

5.4 Linear algebra

In this section, we consider examples of manipulations with vectors and matrices, as well as solving systems of linear algebraic equations and finding eigenvalues with eigenvectors using GSL.

5.4.1 Basic structures

Blocks, vectors, and matrices are the basic structures used to solve systems of linear algebraic equations as well as to find eigenvalues and eigenvectors.

One of the basic structures is a **block**, which is declared in the header file `gsl_block.h`. To create a block, we employ the function

```
gsl_block *b = gsl_block_alloc(100);
```

This function allocates memory for the created block b, which will consist of 100 elements of the `double` type. When we finish working with the block, it must be removed (to free memory):

```
gsl_block_free(b);
```

We illustrate the work with **vectors** on an example of filling a vector with some values.

Listing 5.14.

```
1  #include <stdio.h>
2  #include <gsl/gsl_vector.h>
3  int main(void) {
4      gsl_vector *v = gsl_vector_alloc(20);
5      int i;
6      double a;
7      for (i = 0; i < v->size; i++) {
8          a = i * 3;
9          gsl_vector_set(v, i, a);
```

```
10      }
11      for (i = 0; i < v->size; i++) {
12          printf("a[%d] = %f \n", i, gsl_vector_get(v, i));
13      }
14      gsl_vector_free(v);
15      return 0;
16  }
```

The program output is

```
a[0] = 0.000000
a[1] = 3.000000
a[2] = 6.000000
a[3] = 9.000000
a[4] = 12.000000
```

In this example, we apply the function `gsl_vector_set(v, i, a)` to set the value a to the i-th element of the vector and the function `gsl_vector_get(v, i)` to get the value of the i-th element.

Table 5.6 presents the fundamental vector operations available in GSL.

Table 5.6. Vector operations.

Name	Operation
gsl_vector_add(gsl_vector * a, const gsl_vector * b)	$a_i = a_i + b_i$
gsl_vector_sub(gsl_vector * a, const gsl_vector * b)	$a_i = a_i - b_i$
gsl_vector_mul(gsl_vector * a, const gsl_vector * b)	$a_i = a_i b_i$
gsl_vector_div(gsl_vector * a, const gsl_vector * b)	$a_i = a_i/b_i$
gsl_vector_scale(gsl_vector * a, const double c)	$a_i = c\, a_i$
gsl_vector_add_constant(gsl_vector * a, const double c)	$a_i = c + a_i$
double gsl_vector_max(gsl_vector * a)	$\max_i(a_i)$
double gsl_vector_min(gsl_vector * a)	$\min_i(a_i)$
size_t gsl_vector_max_index(gsl_vector * a)	$i-> \max_i(a_i)$
size_t gsl_vector_min_index(gsl_vector * a)	$i-> \min_i(a_i)$

In GSL, operations with **matrices** are similar to working with vectors. At the beginning, we must allocate memory, and in the end, we must free it. We can access the matrix elements using the functions `gsl_matrix_get(m, i, j)` and `gsl_matrix_set(m, i, j, x)`. The following program demonstrates briefly how to work with matrices.

Listing 5.15.

```
1  #include <stdio.h>
2  #include <gsl/gsl_matrix.h>
3  int main(void) {
4      gsl_matrix *m = gsl_matrix_alloc(3, 3);
5      int i, j;
6      for(i=0; i < m->size1; i++){
7          for(j =0; j < m->size2; j++){
8              gsl_matrix_set(m, i, j, 3*i + 7*j);
9          }
10     }
11     for(i=0; i < m->size1; i++){
12         for(j =0; j < m->size2; j++){
13             printf("a[%d, %d] = %f \n", i, j,
14                     gsl_matrix_get(m, i, j) );
15         }
16     }
17     // The filled vector can be written into the file:
18     FILE *f = fopen("test.txt", "w");
19     gsl_matrix_fprintf(f, m, "%.5g");
20     fclose(f);
21     gsl_matrix_free(m);
22     return 0;
23 }
```

The output is as follow:

```
a[0, 0] = 0.000000
a[0, 1] = 7.000000
a[0, 2] = 14.000000
a[1, 0] = 3.000000
a[1, 1] = 10.000000
a[1, 2] = 17.000000
a[2, 0] = 6.000000
a[2, 1] = 13.000000
a[2, 2] = 20.000000
```

The fundamental matrix operation are shown in Table 5.7.

5.4.2 BLAS support

The **Basic Linear Algebra Subprograms (BLAS)** provide fundamental operations on vectors and matrices, which can be used to optimize algorithms of linear algebra at a higher level. The functionality of BLAS is divided into three levels:

Table 5.7. Matrix operations.

Name	Operation
gsl_matrix_add(gsl_matrix * a, const gsl_matrix * b)	$a_{i,j} = a_{i,j} + b_{i,j}$
gsl_matrix_sub(gsl_matrix * a, const gsl_matrix * b)	$a_{i,j} = a_{i,j} - b_{i,j}$
gsl_matrix_mul_elements(gsl_matrix * a, const gsl_matrix * b)	$a_{i,j} = a_{i,j} \, b_{i,j}$
gsl_matrix_div_elements(gsl_matrix * a, const gsl_matrix * b)	$a_{i,j} = a_{i,j}/b_{i,j}$
gsl_matrix_scale(gsl_matrix * a, const double c)	$a_{i,j} = c \, a_{i,j}$
gsl_matrix_add_constant(gsl_matrix * a, const double c)	$a_{i,j} = c + a_{i,j}$
double gsl_matrix_max(gsl_matrix * a)	$\max_{i,j}(a_{i,j})$
double gsl_matrix_min(gsl_matrix * a)	$\min_{i,j}(a_{i,j})$
gsl_matrix_max_index(gsl_matrix * a, size_t *imax, size_t *jmax)	$(i,j)-> \max_{i,j}(a_{i,j})$
gsl_matrix_min_index(gsl_matrix * a, size_t *imin, size_t *jmin)	$(i,j)-> \min_{i,j}(a_{i,j})$

– *Level 1.* Vector operations, e.g.

$$\mathbf{y} = \alpha \mathbf{x} + \mathbf{y};$$

– *Level 2.* Matrix-vector operations like

$$\mathbf{y} = \alpha A \mathbf{x} + \beta \mathbf{y};$$

– *Level 3.* Matrix-matrix operations, such as

$$C = \alpha A B + C.$$

We demonstrate using Level 1 of BLAS on the following example.

Listing 5.16.

```
1  #include <stdio.h>
2  #include <gsl/gsl_blas.h>
3  int main(void) {
4      int N = 5;
5      gsl_vector *x = gsl_vector_alloc(N);
6      gsl_vector *y = gsl_vector_alloc(N);
7      int i;
8      for(i=0; i < N; i++){
9          gsl_vector_set(x, i, 3*i);
10         gsl_vector_set(y, i, 7*i);
11         printf("x[%d]=%f, y[%d]=%f \n", i, gsl_vector_get(x, i),
12                 i, gsl_vector_get(y, i));
13     }
14     // Calculate the scalar product =(x, y) = x^T * y
```

```
15      double result;
16      gsl_blas_ddot(x, y, &result);
17      printf("\n scalar product = %f \n", result);
18      gsl_vector_free(x);
19      gsl_vector_free(y);
20      return 0;
21  }
```

The program output is as follows:

```
x[0]=0.000000,  y[0]=0.000000
x[1]=3.000000,  y[1]=7.000000
x[2]=6.000000,  y[2]=14.000000
x[3]=9.000000,  y[3]=21.000000
x[4]=12.000000, y[4]=28.000000

scalar product = 630.000000
```

In this example, we use the function `gsl_blas_ddot(x, y, &result)`, which allows calculation of the scalar product for two vectors.

Level 1 of BLAS also has functions for calculating norms $||x|| = \sqrt{\sum_i x_i^2}$ and the absolute sum of the elements of the vector $\sum_i |x_i|$:

Listing 5.17.

```
1  // Computing the Euclidean norm
2  double norm2 = gsl_blas_dnrm2(x);
3  // Computing the absolute sum of the elements of the vector x
4  double sum = gsl_blas_dasum(x);
```

To evaluate vectors according to the formulas $x = a * x$ and $y = a * x + y$, we employ the following functions:

Listing 5.18.

```
1  // Compute   x = a * x
2  gsl_blas_dscal(4.0, x);
3  // Compute   y = a * x + y
4  gsl_blas_daxpy(3.1, x, y);
```

As the result, for the vector *x* and *y* from our example, we get the following values:

```
Compute x=a*x
x[0]=0.000000
x[1]=12.000000
x[2]=24.000000
x[3]=36.000000
x[4]=48.000000

Compute y=a*x+y
y[0]=0.000000
y[1]=44.200000
y[2]=88.400000
y[3]=132.600000
y[4]=176.800000
```

Note that the considered functions also have the implementation for both complex numbers and single-precision numbers (`float`).

Let us consider the function `gsl_blas_dgemv(CblasNoTrans, a, A, x, b, y)` which implements the matrix-vector operation $y = a A x + b y$.

Listing 5.19.

```
1  #include <stdio.h>
2  #include <gsl/gsl_blas.h>
3  int main(void) {
4      int N = 3;
5      gsl_vector *x = gsl_vector_alloc(N);
6      gsl_vector *y = gsl_vector_alloc(N);
7      gsl_matrix *A = gsl_matrix_alloc(N, N);
8      int i, j;
9      for (i = 0; i < N; i++) {
10         gsl_vector_set(x, i, 3*i);
11         gsl_vector_set(y, i, 7*i);
12     }
13     for (i = 0; i < N; i++)
14         for(j = 0; j < N; j++)
15             gsl_matrix_set(A, i, j, i*j);
16     // Compute sum y = a * op(A)*x + b*y
17     double a = 9;
18     double b = 2.7;
19     gsl_blas_dgemv(CblasNoTrans, a, A, x, b, y);
20     for (i = 0; i < N; i++){
21         printf("y[%d]=%f \n", i, gsl_vector_get(y, i));
22     }
23     gsl_vector_free(x);
24     gsl_vector_free(y);
```

```
25      gsl_matrix_free(A);
26      return 0;
27  }
```

The output is as follows:

```
y[0]=0.000000
y[1]=153.900000
y[2]=307.800000
```

Level 2 of BLAS has other functions for matrix-vector operations, such as $x = A x$, $A = a x y^T + A$, as well as functions for some particular matrices (symmetric, triangular, hermitian etc.).

Now apply the last level (Level 3) of BLAS associated with matrix–matrix operations. The following example illustrates the operation of the type $C = a A B + b C$.

Listing 5.20.

```
1   #include <stdio.h>
2   #include <gsl/gsl_blas.h>
3   int main(void) {
4       int N = 3;
5       gsl_matrix *A = gsl_matrix_alloc(N, N);
6       gsl_matrix *B = gsl_matrix_alloc(N, N);
7       gsl_matrix *C = gsl_matrix_alloc(N, N);
8       int i, j;
9       for (i = 0; i < N; i++) {
10          for (j = 0; j < N; j++) {
11              gsl_matrix_set(A, i, j, i*j+9);
12              gsl_matrix_set(B, i, j, i+j);
13              gsl_matrix_set(C, i, j, i*2/(j+1));
14          }
15      }
16      // Compute C = a * op(A) * op(B) + b * C, op(A) = A, AT, AH
17      double a = 9;
18      double b = 2.7;
19      gsl_blas_dgemm(CblasNoTrans, CblasNoTrans, a, A, B, b, C);
20      for (i = 0; i < N; i++)
21          for (j = 0; j < N; j++)
22              printf("C[%d, %d]=%f \n", i, j,
23                      gsl_matrix_get(C, i, j));
24      gsl_matrix_free(A);
25      gsl_matrix_free(B);
26      gsl_matrix_free(C);
27      return 0;
28  }
```

The result is as follows:

```
C[0, 0]=243.000000
C[0, 1]=486.000000
C[0, 2]=729.000000
C[1, 0]=293.400000
C[1, 1]=560.700000
C[1, 2]=828.000000
C[2, 0]=343.800000
C[2, 1]=635.400000
C[2, 2]=929.700000
```

There are also other matrix–matrix operations, which can be found in the GSL Reference Manual.

5.4.3 Solving linear systems

The GSL library provides a series of functions for numerical solving of linear systems

$$Ax = b$$

via direct methods (*LU*, *QR*-decomposition, and others).

As an example, let us consider the process of solving a linear system by means of the *LU* decomposition

$$PA = LU,$$

where *P* is a permutation matrix, *L* stands for a lower triangular matrix, and *U* denotes an upper triangular matrix. Such a decomposition lets us solve the linear system in two steps:

$$Ly = Pb,$$

$$Ux = y.$$

Since *L* is a lower triangular matrix, then at the first step the system is solved by the forward substitution. In the second step, since *U* is an upper triangular matrix, and the system is solved by backwards substitution.

The application of the *LU* **decomposition** is as follows.

Listing 5.21.

```
1  #include <stdio.h>
2  #include <gsl/gsl_linalg.h>
3  int main(void) {
4      int N = 4;
5      double a_data[] = {2,  1,  0,  0,
```

```
6                              2,   3,  -1,   0,
7                              0,   1,  -1,   3,
8                              0,   0,   1,  -1};
9       double b_data[] = {4,   9, 12,  -4};
10      gsl_matrix_view A = gsl_matrix_view_array(a_data, N, N);
11      gsl_vector_view b = gsl_vector_view_array(b_data, N);
12      gsl_vector      *x = gsl_vector_alloc(N);
13      gsl_permutation *p = gsl_permutation_alloc(N);
14      int signum; // the sign of the permutation
15      gsl_linalg_LU_decomp(&A.matrix, p, &signum);
16      gsl_linalg_LU_solve(&A.matrix, p, &b.vector, x);
17      printf("Solution:\n");
18      gsl_vector_fprintf(stdout, x, "%g");
19      gsl_vector_free(x);
20      gsl_permutation_free(p);
21      return 0;
22  }
```

The obtained result is as follows:

```
Solution:
1
2
-1
3
```

In this example, we use the function

```
gsl_linlalg_LU_decomp(&A.matrix, p, &signum)
```

which executes *LU* decomposition of the matrix *A*, and the function

```
gsl_linalg_LU_solve(&A.matrix, p, &b.vector, x)
```

which solves the linear system on the basis of obtained *LU* decomposition.

It should be noted that on the basis of the *LU* decomposition we can also deter-mine the determinant of a matrix using the function

```
gsl_linalg_LU_det(&A.matrix, signum)
```

as well as invert a matrix using the function

```
gsl_linalg_LU_invert(&A.matrix, p, Ainv)
```

where `Ainv` is the derived inverse matrix.

To solve linear systems, we may use the *QR* **decomposition,** which is the decomposition of the matrix *A* into an orthogonal matrix Q ($QQ^T = I$) and a right-triangular matrix *R* using the functions

```
gsl_vector *tau = gsl_vector_alloc(N);
gsl_linalg_QR_decomp(&A.matrix, tau);
gsl_linalg_QR_solve(&A.matrix, tau, &b.vector, x);
```

The library has some modifications of these methods as well as implementations for symmetric matrices *A*.

5.4.4 Eigenvalues and eigenvectors

The GSL library contains functions to evaluate **eigenvalues and eigenvectors** of matrices. These functions are declared in the header file `gsl_eigen.h`. The following program calculates eigenvalues and eigenvectors of the symmetric matrix H: $H_{i,j} = 2 * i + 3 * j$.

Listing 5.22.

```
1  #include <stdio.h>
2  #include <gsl/gsl_math.h>
3  #include <gsl/gsl_eigen.h>
4  int main(void) {
5      int N = 4;
6      int i, j;
7      // Matrix initialization
8      gsl_matrix *H = gsl_matrix_alloc(N, N);
9      for (i = 0; i < N; i++)
10          for (j = 0; j < N; j++)
11              gsl_matrix_set(H, i, j, 2*i + 3*j));
12      // Calculate the eigenvalues and eigenvectors
13      gsl_matrix *E = gsl_matrix_alloc(N, N);
14      gsl_vector *v = gsl_vector_alloc(N);
15      gsl_eigen_symmv_workspace *w = gsl_eigen_symmv_alloc(N);
16      gsl_eigen_symmv(H, v, E, w);
17      gsl_eigen_symmv_free(w);
18      // Sorting
19      gsl_eigen_symmv_sort(v, E, GSL_EIGEN_SORT_ABS_ASC);
```

```
20    // Print
21    for (i = 0; i < N; i++) {
22        double evalue = gsl_vector_get(v, i);
23        gsl_vector_view evector = gsl_matrix_column(E, i);
24
25        printf("Eigen Value[%d]=%f\n", i, evalue);
26        printf("Eigen Vector=[");
27        for (j = 0; j < N; j++)
28            printf("%f ", gsl_vector_get(&evector.vector, j));
29        printf("]\n\n");
30    }
31    gsl_matrix_free(H);
32    gsl_matrix_free(E);
33    gsl_vector_free(v);
34    return 0;
35 }
```

The output is as follows:

```
Eigen Value[0]=0.302921
Eigen Vector=[-0.181557 0.666995 -0.686333 0.226057 ]

Eigen Value[1]=0.503409
Eigen Vector=[-0.461194 0.536010 0.468452 -0.529666 ]

Eigen Value[2]=-2.484143
Eigen Vector=[0.838398 0.331894 -0.041005 -0.430411 ]

Eigen Value[3]=31.677813
Eigen Vector=[0.226774 0.397061 0.554814 0.695052 ]
```

To calculate eigenvalues and eigenvectors of the matrix, we must allocate memory:

Listing 5.23.

```
14  gsl_eigen_symmv_workspace *w = gsl_eigen_symmv_alloc(N);
```

Then we call the function for calculating eigenvalues and eigenvectors:

Listing 5.24.

```
12  gsl_matrix *E = gsl_matrix_alloc(N, N);
13  gsl_vector *v = gsl_vector_alloc(N);
14  . . .
15  gsl_eigen_symmv(H, v, E, w);
```

and free memory at the end:

Listing 5.25.

```
16  gsl_eigen_symmv_free(w);
```

If necessary we can sort the obtained values:

Listing 5.26.

```
18  gsl_eigen_symmv_sort(v, E, GSL_EIGEN_SORT_ABS_ASC);
```

The library has functions for obtaining eigenvalues and eigenvectors of nonsymmetric matrices:

```
gsl_matrix_complex *E = gsl_matrix_complex_alloc(N, N);
gsl_vector_complex *v = gsl_vector_complex_alloc(N);
gsl_eigen_nonsymmv_workspace *w = gsl_eigen_nonsymmv_alloc(N);
gsl_eigen_nonsymmv(H, v, E, w);
gsl_eigen_nonsymmv_free(w);
gsl_eigen_nonsymmv_sort(v, E, GSL_EIGEN_SORT_ABS_ASC);
```

Note that in the case of nonsymmetric matrices, eigenvalues and eigenvectors are complex numbers.

5.5 Numerical differentiation and integration

We consider examples of numerical differentiation of functions and approximate calculation of a definite integral using GSL.

5.5.1 Numerical differentiation

In the GSL library, the numerical calculation of derivatives of a function is performed using the finite difference method. There are three types of finite differences: central, forward, and backward.

Assume that we need to evaluate the **numerical derivative** of the function

$$f(x) = x^{\frac{5}{4}} + \sin(x),$$

at the points $x = 5$ and $x = 0$ and compare it with the exact analytical derivative

$$f'(x) = \frac{5}{4}x^{\frac{1}{4}} + \cos(x).$$

The function $f(x)$ does not have real values for $x < 0$, and the derivative at the point $x = 0$ must therefore be evaluated using the forward difference:

Listing 5.27.

```
1  #include <stdio.h>
2  #include <gsl/gsl_deriv.h>
3  double f(double x, void * params) {
4      return pow(x, 1.25) + sin(x);
5  }
6  int main(void) {
7      // Define the integrand
8      gsl_function F;
9      F.function = &f;
10     F.params = 0;
11     // Define the step of difference derivative
12     double h = 1e-3;
13     // Define variables
14     double result, abserr;
15     // Set the first point
16     double x = 5;
17     // Calculate the derivative using the central difference
18     gsl_deriv_central(&F, x, h, &result, &abserr);
19     printf("f'(%.0f) = %.10f +/- %.10f\n", x, result, abserr);
20     printf("exact = %.10f\n", 1.25 * pow(x, 0.25) + cos(x));
21     // Set the second point
22     x = 0;
23     // Calculate the derivative using the forward difference
24     gsl_deriv_forward(&F, x, h, &result, &abserr);
25     // Print the result and compare with the exact derivative
26     printf("f'(%.0f) = %.10f +/- %.10f\n", x, result, abserr);
27     printf("exact = %.10f\n", 1.25 * pow(x, 0.25) + cos(x));
28     return 0;
29  }
```

We start with including the header file to use functions for numerical differentiation:

Listing 5.28.

```
2  #include <gsl/gsl_deriv.h>
```

The function $f(x)$ must be created as `gsl_function`:

Listing 5.29.

```
8   gsl_function F;
9   F.function = &f;
10  F.params = 0;
```

where `&f` is the pointer to the C function:

Listing 5.30.

```
3   double f(double x, void * params) {
4       return pow(x, 1.25) + sin(x);
5   }
```

Further, we define the variables for the calculated derivative and its absolute error. Then we call the function for evaluating the approximate derivative using the central difference:

Listing 5.31.

```
14  double result, abserr;
15  . . .
16  double x = 5;
17  . . .
18  gsl_deriv_central(&F, x, h, &result, &abserr);
```

To call the function, it is necessary to specify as arguments the differentiable function, the point, and the step-size. The function returns the calculated derivative and its absolute error. We compare the obtained result with the analytical value:

Listing 5.32.

```
19  printf("f'(%.0f) = %.10f +/- %.10f\n", x, result, abserr);
20  printf("exact = %.10f\n", 1.25 * pow(x, 0.25) + cos(x));
```

Similarly, we calculate the derivative at the point $x = 0$ using the forward difference:

Listing 5.33.

```
22  x = 0;
23  . . .
24  gsl_deriv_forward(&F, x, h, &result, &abserr);
25  . . .
26  printf("f'(%.0f) = %.10f +/- %.10f\n", x, result, abserr);
27  printf("exact = %.10f\n", 1.25 * pow(x, 0.25) + cos(x));
```

The output of the program is as follows:

```
f'(5) = 2.1528481619 +/- 0.0000000003
exact = 2.1528481620
f'(0) = 1.0034436513 +/- 0.0022299192
exact = 1.0000000000
```

The library also has a function to calculate the approximate derivative using backward difference `gsl_deriv_backward`.

5.5.2 Numerical integration

Here we shall discuss the numerical methods for integration available in the library, such as calculating definite integrals over finite or infinite intervals, integrals of functions with singularities or periodic functions, and so on.

Algorithms of **numerical integration** calculate an approximate value of a definite integral of the form

$$I = \int_a^b f(x) \, w(x) \, dx,$$

where $w(x)$ is a weight function (in particular, $w(x) = 1$).

The names of the methods of numerical integration are based on the first letters of the methods (see Table 5.8).

In the following is a short list containing some methods of numerical integration:

- QNG – a nonadaptive Gauss–Kronrod integration, $w(x) = 1$;
- QAG – a simple adaptive integration;
- QAGP – adaptive integration with known singular points;
- QAWO – adaptive integration for oscillatory functions $w(x) = \cos(\omega x)$ or $w(x) = \sin(\omega x)$.

Table 5.8. The letters in the name of methods for numerical integration.

#	Letter	Description
1	Q	quadrature
2	N	non-adaptive
	A	adaptive
3	G	general
	W	weight function
	without letter	Simple
	S	singularities
	P	points of special difficulty
4	I	infinite range
	O	oscillatory
	F	Fourier transform
	C	Cauchy principal value

Consider the numerical integration of the function with an algebraic-logarithmic singularity at the origin:

$$\int_0^1 \frac{log(x)}{x^{\frac{1}{3}}} dx = -2,25.$$

The program is as follows:

Listing 5.34.

```
1   #include <stdio.h>
2   #include <gsl/gsl_integration.h>
3   double f(double x, void * params) {
4       return log(x) / pow(x, 1.0 / 3.0);
5   }
6   int main(void) {
7       // Define the integrand
8       gsl_function F;
9       F.function = &f;
10      F.params = 0;
11      // Define the interval of integration
12      double a = 0;
13      double b = 1;
14      // Set the absolute and relative errors
15      double abserr = 0.0;
16      double relerr = 1e-3;
17      // Set the maximum number of additional subintervals
18      int max_intervals = 10;
19      // Allocate memory for the additional subintervals
20      gsl_integration_workspace * w =
21          gsl_integration_workspace_alloc(max_intervals);
```

```
22      // Variable for the results
23      double result, error;
24      // Numerical integration using QAGS algorithm
25      gsl_integration_qags(&F, a, b, abserr, relerr, max_intervals,
26                           w, &result, &error);
27      // Free memory
28      gsl_integration_workspace_free(w);
29      // Print the results
30      printf("result           = % .20f\n", result);
31      printf("exact result     = % .20f\n", -2.25);
32      printf("estimated error = % .20f\n", error);
33      printf("actual error     = % .20f\n", result + 2.25);
34      printf("intervals =  %d\n", w->size);
35      return 0;
36  }
```

It is necessary to include the header file in order to use the functions of numerical integration:

Listing 5.35.

```
2  #include <gsl/gsl_integration.h>
```

Next, we define the integrand as the structure `gsl_function`:

Listing 5.36.

```
8   gsl_function F;
9   F.function = &f;
10  F.params = 0;
```

where `&f` is a pointer to the C function that describes the integrand. Further, we define the interval along with the absolute and relative error limits:

Listing 5.37.

```
12  double a = 0;
13  double b = 1;
14  . . .
15  double abserr = 0.0;
16  double relerr = 1e-3;
```

If the absolute error limit `abserr` equals zero, only the relative error `relerr` will be considered.

Since the integrand has singularity at the point $x = 0$, it is necessary to apply the QAGS algorithm. To employ adaptive integration, we must define the maximum number of subintervals and allocate memory for them:

Listing 5.38.

```
18  int max_intervals = 10;
19  . . .
20  gsl_integration_workspace * w =
21      gsl_integration_workspace_alloc(max_intervals);
```

Further, we call the numerical integration function:

Listing 5.39.

```
23  double result, error;
24  . . .
25  gsl_integration_qags(&F, a, b, abserr, relerr, max_intervals,
26                       w, &result, &error);
```

At the end, we free memory and print the results:

Listing 5.40.

```
28  gsl_integration_workspace_free(w);
29  . . .
30  printf("result           = % .20f\n", result);
31  printf("exact result     = % .20f\n", -2.25);
32  printf("estimated error = % .20f\n", error);
33  printf("actual error     = % .20f\n", result + 2.25);
34  printf("intervals =  %d\n", w->size);
```

The output of the above program is as follows:

```
result           = -2.25000000000004529710
exact result     = -2.25000000000000000000
estimated error =  0.00117996545077936332
actual error     = -0.00000000000004529710
intervals =  6
```

The error returned by the QAGS function is larger than the actual error. The defined relative error will be achieved with six additional subintervals of the integration.

GSL involves other methods of numerical integration; see the documentation for more details.

5.5.3 Numerical integration of multidimensional functions

For the numerical integration of multidimensional functions in GSL, *Monte Carlo's* methods are employed. Using these methods, we can calculate integrals of the form

$$I = \int_{a_1}^{b_1} \int_{a_2}^{b_2} \cdots \int_{a_n}^{b_n} f(x_1, x_2, \ldots, x_n) \, dx_1 \, dx_2 \cdots dx_n.$$

Three **Monte Carlo integration methods** are implemented in the library, i.e.
-   PLAIN is a simple Monte Carlo method, where integration points are randomly defined;
-   MISER is a method with integration points concentrated in regions of the highest variance;
-   VEGAS is a method similar to MISER, but it has a different logic of concentrating integration points.

Let us integrate numerically the following three-dimensional function:

$$I = \int_0^\pi \int_0^\pi \int_0^\pi \frac{dx_1 \, dx_2 \, dx_3}{1 - \cos(x_1) \cos(x_2) \cos(x_3)} = \frac{1}{4} \Gamma^4 \left(\frac{1}{4} \right),$$

where $\Gamma(x)$ is the gamma function defined in the header file gsl_math.h. The corresponding program is as follows:

Listing 5.41.

```
1  #include <stdlib.h>
2  #include <gsl/gsl_math.h>
3  #include <gsl/gsl_sf.h>
4  #include <gsl/gsl_monte_plain.h>
5  double f(double *x, size_t dim, void *params) {
6      return 1.0 / (1.0 - cos (x[0]) * cos (x[1]) * cos (x[2]));
7  }
8  int main() {
9      // Define integrand
10     gsl_monte_function F = {&f, 3, 0};
11     // Define the integration region
12     double a[3] = {0, 0, 0};
13     double b[3] = {M_PI, M_PI, M_PI};
14     // Number of integration points
15     size_t calls = 10000000;
```

```
16     // Prepare random numbers generator
17     gsl_rng_env_setup();
18     gsl_rng *rand = gsl_rng_alloc(gsl_rng_default);
19     // Variable for the results
20     double result, error;
21     // Allocate memory and integrate using Monte Carlo method
22     gsl_monte_plain_state *state = gsl_monte_plain_alloc(3);
23     gsl_monte_plain_integrate(&F, a, b, 3, calls, rand, state,
24         &result, &error);
25     gsl_monte_plain_free(state);
26     // Free memory of the random number generator
27     gsl_rng_free(rand);
28     // Exact value of the integral
29     double exact = pow(gsl_sf_gamma(0.25), 4) / 4;
30     // Print result and compare with exact solution
31     printf ("result       = %.10f\n", result);
32     printf ("exact result = %.10f\n", exact);
33     printf ("error        = %.10f\n", error);
34     printf ("exact error  = %.10f\n", exact - result);
35     return 0;
36  }
```

First, we include the header file of the standard Monte Carlo interpolation method:

Listing 5.42.

```
4   #include <gsl/gsl_monte_plain.h>
```

The integrand is defined as the structure `gsl_monte_function`:

Listing 5.43.

```
10  gsl_monte_function F = {&f, 3, 0};
```

where `&f` is a reference to the integrand

Listing 5.44.

```
6   double f(double *x, size_t dim, void *params) {
7       return 1.0 / (1.0 - cos (x[0]) * cos (x[1]) * cos (x[2]));
8   }
```

Here `*x` are coordinates of a point, `dim` is the dimension, and `*params` is a pointer to parameters.

Next, we define the integration region and number of integration points:

Listing 5.45.

```
12  double a[3] = {0, 0, 0};
13  double b[3] = {M_PI, M_PI, M_PI};
14  . . .
15  size_t calls = 10000000;
```

Since the Monte Carlo method uses random points, we must create the random numbers generator. Apply the default generator:

Listing 5.46.

```
17  gsl_rng_env_setup();
18  gsl_rng *rand = gsl_rng_alloc(gsl_rng_default);
```

All input parameters for integration of the function are now ready. Now we declare the variables for the results, allocate memory, and call the function of the standard Monte Carlo method:

Listing 5.47.

```
20  double result, error;
21  . . .
22  gsl_monte_plain_state *state = gsl_monte_plain_alloc(3);
23  gsl_monte_plain_integrate(&F, a, b, 3, calls, rand, state,
24      &result, &error);
```

Further, we must free memory:

Listing 5.48.

```
25  gsl_monte_plain_free(state);
26  . . .
27  gsl_rng_free(rand);
```

Finally, we calculate the exact solution using the special function $\Gamma(x)$ and compare it with the obtained result:

Listing 5.49.

```
29  double exact = pow(gsl_sf_gamma(0.25), 4) / 4;
30  . . .
31  printf ("result        = %.10f\n", result);
32  printf ("exact result = %.10f\n", exact);
33  printf ("error         = %.10f\n", error);
34  printf ("exact error  = %.10f\n", exact - result);
```

After compilation and execution of the program, we get the following results:

```
result       = 43.1275381572
exact result = 43.1980665159
error        = 0.0749989525
exact error  = 0.0705283587
```

To use MISER and VEGAS methods, we must include `gsl_monte_miser.h` and `gsl_monte_vegas.h`. In this case, the general principle of the implementation of Monte Carlo methods is the same.

5.6 Cauchy problem for systems of ODEs

Now using the GSL library, we solve the **Cauchy problem for systems of ordinary differential equations (ODEs)**

$$\frac{du_i(t)}{dt} = f_i(t, u_1(t), u_2(t), \ldots, u_n(t)), \quad i = 1, 2, \ldots, n, \quad 0 < t \le T,$$

$$u_i(0) = u_i^0, \quad i = 1, 2, \ldots, n.$$

To solve ODE systems, it is possible to apply the following functions of the GSL library:
- *the stepping functions*, which compute the next time level using a fixed time step;
- *the evolution functions*, which compute the next time level using the optimal time step;
- *the driver object*, which computes the next time level hiding solution process.

5.6.1 Stepping functions

These functions of calculating the solution at the next time level are fundamental components, and are used in other methods for solving ODE systems. The library provides **explicit, implicit, and multistep methods**, presented in Table 5.9. Implicit and multistep methods are used only with control functions, except the implicit Bulirsch–Stoer method of Bader and Deuflhard.

Table 5.9. Stepping functions.

Name	Description
gsl_odeiv2_step_rk2	Explicit embedded Runge–Kutta (2, 3) method
gsl_odeiv2_step_rk4	Explicit 4th order (classical) Runge–Kutta
gsl_odeiv2_step_rkf45	Explicit embedded Runge-Kutta–Fehlberg (4, 5) method
gsl_odeiv2_step_rkck	Explicit embedded Runge-Kutta Cash–Karp (4, 5) method
gsl_odeiv2_step_rk8pd	Explicit embedded Runge-Kutta Prince–Dormand (8, 9) method
gsl_odeiv2_step_rk1imp	Implicit Gaussian first order Runge–Kutta
gsl_odeiv2_step_rk2imp	Implicit Gaussian second order Runge–Kutta
gsl_odeiv2_step_rk4imp	Implicit Gaussian 4th order Runge–Kutta
gsl_odeiv2_step_bsimp	Implicit Bulirsch-Stoer method of Bader and Deuflhard
gsl_odeiv2_step_msadams	Linear multistep Adams method in Nordsieck form
gsl_odeiv2_step_msbdf	Linear multistep backward differentiation formula method in Nordsieck form

Consider the Cauchy problem for the ordinary differential equation:

$$\frac{du}{dt} = f(t, u), \quad 0 < t \leq \pi,$$
$$f(t, u) = u - \cos(t) + \sin(t),$$
$$u(0) = 0,$$

with the exact solution $u(t) = \sin(t)$. The program for the numerical solution of this problem is as follows:

Listing 5.50.

```
1  #include <stdio.h>
2  #include <gsl/gsl_math.h>
3  #include <gsl/gsl_errno.h>
4  #include <gsl/gsl_odeiv2.h>
5  int F(double t, const double u[], double f[], void *params) {
6      f[0] = u[0] - sin(t) + cos(t);
7      return GSL_SUCCESS;
8  }
9  int main(void) {
10     // Set the ODE system
11     gsl_odeiv2_system system = {F, NULL, 1, NULL};
12     // Time step parameters
13     double tau = 0.1;
```

```
14    double t = 0.0;
15    double T = M_PI;
16    // Select the explicit 4th order Runge-Kutta method
17    const gsl_odeiv2_step_type* st = gsl_odeiv2_step_rk4;
18    // Create the stepping function
19    gsl_odeiv2_step* stepping = gsl_odeiv2_step_alloc(st, 1);
20    // Define the unknown and set the initial value
21    double u[1] = {0.0};
22    double err[1] = {0.0};
23    // Apply the stepping function
24    while (t < T) {
25        gsl_odeiv2_step_apply(stepping, t, tau, u, err,
26                              NULL, NULL, &system);
27        t = t + tau;
28    }
29    // Free memory
30    gsl_odeiv2_step_free(stepping);
31    // Print the results
32    printf("t = %.1f\nresult = %.10f\nerror = %.10f\n",
33           t, u[0], err[0]);
34    printf("exact error = %.10f\n", sin(t) - u[0]);
35    return 0;
36 }
```

As before, we include first the header file:

Listing 5.51.

```
4  #include <gsl/gsl_odeiv2.h>
```

This is a new implementation of the code for solving ODEs (variant 2). The old implementation with the header file gsl_odeiv.h is retained for backwards compatibility.

To define an ODE system, GSL uses the special structure gsl_odeiv2_system:

Listing 5.52.

```
11  gsl_odeiv2_system system = {F, NULL, 1, NULL};
```

The elements of the structure are a pointer to the function $f_i(t, u_1, u_2, \ldots, u_n)$, a pointer to the Jacobian matrix, a dimension of the system, and a pointer to the parameters of the system. The function F has the following form:

Listing 5.53.

```
5  int F(double t, const double u[], double f[], void *params) {
6      f[0] = u[0] - sin(t) + cos(t);
7      return GSL_SUCCESS;
8  }
```

Further, we set the time step and range:

Listing 5.54.

```
13  double tau = 0.1;
14  double t = 0.0;
15  double T = M_PI;
```

Since we consider the simplest approach, the explicit 4th-order Runge–Kutta method is chosen to calculate the solution at the next time level:

Listing 5.55.

```
17  const gsl_odeiv2_step_type* st = gsl_odeiv2_step_rk4;
18  . . .
19  gsl_odeiv2_step* stepping = gsl_odeiv2_step_alloc(st, 1);
```

Next, we define the variables for the solution and error and set the time loop to evaluate the solution:

Listing 5.56.

```
21  double u[1] = {0.0};
22  double err[1] = {0.0};
23  . . .
24  while (t < T) {
25      gsl_odeiv2_step_apply(stepping, t, tau, u, err,
26                            NULL, NULL, &system);
27      t = t + tau;
28  }
```

The function `gsl_odeiv2_step_apply` gets 6 mandatory and 2 optional parameters.

The optional parameters are an input array containing the derivatives ($\frac{du_i}{dt}$) for the current time level t and an output array with computed derivatives for the next time level $t + \tau$.

These parameters are prescribed to avoid repeating the computation of the derivatives $(\frac{du_i}{dt})$.

Finally, we free memory and print the results:

Listing 5.57.

```
30  gsl_odeiv2_step_free(stepping);
31  . . .
32  printf("t = %.1f\nresult = %.10f\nerror = %.10f\n",
33          t, u[0], err[0]);
34  printf("exact error = %.10f\n", sin(t) - u[0]);
```

The output of the program is as follows:

```
t = 3.2
result = -0.0583737251
error = -0.0000000085
exact error = -0.0000004183
```

5.6.2 Evolution function

If we apply explicit methods, then we should choose a time step so that the solution does not diverge. For this, the time step is selected to be sufficiently small. Therefore, the number of time levels may essentially increase. The **evolution function** is used to automatically select the optimal time step for advancing the solution forward one step.

Let us consider the Cauchy problem for the following system of ODEs:

$$\frac{du_1}{dt} = u_2, \quad 0 < t \le \pi,$$

$$\frac{du_2}{dt} = -u_1, \quad 0 < t \le \pi,$$

$$u_1(0) = 0,$$

$$u_2(0) = 1,$$

with the exact solution $u_1(t) = \sin(t), u_2(t) = \cos(t)$. Prepare the following program:

Listing 5.58.

```
1  #include <stdio.h>
2  #include <gsl/gsl_math.h>
3  #include <gsl/gsl_errno.h>
```

```c
 4  #include <gsl/gsl_odeiv2.h>
 5  int F(double t, const double u[], double f[], void *params) {
 6      f[0] = u[1];
 7      f[1] = -u[0];
 8      return GSL_SUCCESS;
 9  }
10  int main(void) {
11      // Set the system of equations
12      gsl_odeiv2_system system = {F, NULL, 2, NULL};
13      // Time step parameters
14      double tau = 0.1;
15      double t = 0.0;
16      double T = M_PI;
17      // Select the explicit 4-th order Runge-Kutta method
18      const gsl_odeiv2_step_type* st = gsl_odeiv2_step_rk4;
19      // Create the stepping function
20      gsl_odeiv2_step* stepping = gsl_odeiv2_step_alloc(st, 2);
21      // Create the control object
22      double abserr = 1e-6;
23      double relerr = 0.0;
24      gsl_odeiv2_control* control =
25          gsl_odeiv2_control_y_new(abserr, relerr);
26      // Create the evolution function
27      gsl_odeiv2_evolve* evolve = gsl_odeiv2_evolve_alloc(2);
28      // Define the unknown and set initial value
29      double u[2] = {0.0, 1.0};
30      int steps = 0;
31      // Apply the evolution function
32      while (t < T) {
33          gsl_odeiv2_evolve_apply(evolve, control, stepping,
34              &system, &t, T, &tau, u);
35          steps++;
36      }
37      // Free memory
38      gsl_odeiv2_evolve_free(evolve);
39      gsl_odeiv2_control_free(control);
40      gsl_odeiv2_step_free(stepping);
41      // Print the results
42      printf("t = %.1f\tsteps = %d\n", t, steps);
43      printf("result[0] = %.10f\n", u[0]);
44      printf("error[0]  = %.10f\n", sin(t) - u[0]);
45      printf("result[1] = %.10f\n", u[1]);
46      printf("error[1]  = %.10f\n", cos(t) - u[1]);
47      return 0;
48  }
```

Unlike the previous example, we consider the system of two equations, but it does not change the algorithm much. In addition, we employ the control and evolution functions:

Listing 5.59.

```
22  double abserr = 1e-6;
23  double relerr = 0.0;
24  gsl_odeiv2_control* control =
25      gsl_odeiv2_control_y_new(abserr, relerr);
26  . . .
27  gsl_odeiv2_evolve* evolve = gsl_odeiv2_evolve_alloc(2);
```

The selection of the time step depends on the absolute `abserr` and relative `relerr` errors. When we call the evolution function, the selection of the time step and the advance of the solution to the next time level is performed within the function:

Listing 5.60.

```
29  double u[2] = {0.0, 1.0};
30  int steps = 0;
31  . . .
32  while (t < T) {
33      gsl_odeiv2_evolve_apply(evolve, control, stepping,
34          &system, &t, T, &tau, u);
35      steps++;
36  }
```

The output of the program is as follows:

```
t = 3.1 steps = 16
result[0] = 0.0000026904
error[0]  = -0.0000026904
result[1] = -0.9999997717
error[1]  = -0.0000002283
```

After calculating 16 time levels, we obtain the error, which is very close to the presribed absolute error `abserr = 1e-6`.

5.6.3 Driver

GSL provides a special driver object for the easy use of implicit and multistep methods. The application of driver objects illustrates the solution of the following system of ODEs:

$$\frac{du_1}{dt} = u_2, \quad 0 < t \leq \pi,$$

$$\frac{du_2}{dt} = -u_1 + u_2^2 - \cos^2(t), \quad 0 < t \leq \pi,$$

$$u_1(0) = 0,$$

$$u_2(0) = 1,$$

with the exact solution $u_1(t) = \sin(t), u_2(t) = \cos(t)$. To solve this problem, we apply the following code:

Listing 5.61.

```
1  #include <stdio.h>
2  #include <gsl/gsl_math.h>
3  #include <gsl/gsl_matrix.h>
4  #include <gsl/gsl_errno.h>
5  #include <gsl/gsl_odeiv2.h>
6  int F(double t, const double u[], double f[], void *params) {
7      f[0] = u[1];
8      f[1] = - u[0] + u[1] * u[1] - cos(t) * cos(t);
9      return GSL_SUCCESS;
10 }
11 int J(double t, const double u[], double *dfdu, double dfdt[],
12        void *params) {
13     gsl_matrix_view dfdu_mat = gsl_matrix_view_array(dfdu, 2, 2);
14     gsl_matrix* m = &dfdu_mat.matrix;
15     gsl_matrix_set(m, 0, 0, 0.0);
16     gsl_matrix_set(m, 0, 1, 1.0);
17     gsl_matrix_set(m, 1, 0, -1.0);
18     gsl_matrix_set(m, 1, 1, 2 * u[1]);
19     dfdt[0] = 0.0;
20     dfdt[1] = -2 * cos(t) * sin(t);
21     return GSL_SUCCESS;
22 }
23 int main(void) {
24     // Define the  system  of equations
25     gsl_odeiv2_system system = {F, J, 2, NULL};
26     //Time step parameters
27     double tau = 0.1;
28     double t = 0.0;
29     double T = M_PI;
30     // Select the explicit 4-th order Runge-Kutta method
31     const gsl_odeiv2_step_type* st = gsl_odeiv2_step_rk4imp;
32     // Set errors and create the driver object
33     double abserr = 1e-6;
34     double relerr = 0.0;
```

```
35    gsl_odeiv2_driver* driver =
36        gsl_odeiv2_driver_alloc_y_new(&system, st, tau,
37                                      abserr, relerr);
38    // Define the unknown and set the initial value
39    double u[2] = {0.0, 1.0};
40    // Apply the driver object
41    while (t < T) {
42        gsl_odeiv2_driver_apply(driver, &t, t + tau, u);
43    }
44    // Free memory
45    gsl_odeiv2_driver_free(driver);
46    // Print the results
47    printf("t = %.1f\n", t);
48    printf("result[0] = %.10f\n", u[0]);
49    printf("error[0]  = %.10f\n", sin(t) - u[0]);
50    printf("result[1] = %.10f\n", u[1]);
51    printf("error[1]  = %.10f\n", cos(t) - u[1]);
52    return 0;
53  }
```

In contrast to the previous example with the evolution function, in the system of equations

Listing 5.62.

```
25  gsl_odeiv2_system system = {F, J, 2, NULL};
```

we specify the function that stores the Jacobian matrix ($\frac{\partial f_i}{\partial u_j}$) and the vector of the time derivative ($\frac{\partial f_i}{\partial t}$):

Listing 5.63.

```
11  int J(double t, const double u[], double *dfdu, double dfdt[],
12        void *params) {
13      gsl_matrix_view dfdu_mat = gsl_matrix_view_array(dfdu, 2, 2);
14      gsl_matrix* m = &dfdu_mat.matrix;
15      gsl_matrix_set(m, 0, 0, 0.0);
16      gsl_matrix_set(m, 0, 1, 1.0);
17      gsl_matrix_set(m, 1, 0, -1.0);
18      gsl_matrix_set(m, 1, 1, 2 * u[1]);
19      dfdt[0] = 0.0;
20      dfdt[1] = -2 * cos(t) * sin(t);
21      return GSL_SUCCESS;
22  }
```

Here we introduce the special driver object:

Listing 5.64.

```
33  double abserr = 1e-6;
34  double relerr = 0.0;
35  gsl_odeiv2_driver* driver =
36      gsl_odeiv2_driver_alloc_y_new(&system, st, tau,
37                                    abserr, relerr);
```

where we set the type of stepping function gsl_odeiv2_step_type*, the maximal time step tau as well as the absolute abserr and relative relerr errors. The driver object automatically initializes the evolution function, control object, and stepping function.

To evaluate the solution at each time level, we apply the driver object:

Listing 5.65.

```
41  while (t < T) {
42      gsl_odeiv2_driver_apply(driver, &t, t + tau, u);
43  }
```

The output of the program is as follows:

```
t = 3.2
result[0] = -0.0583741821
error[0]  = 0.0000000387
result[1] = -0.9982947792
error[1]  = 0.0000000035
```

Of the three methods of solving systems of ODEs, the latter is the easiest for implementation and provides more accurate results.

5.7 Solution of equations

The GSL library allows to solve equations and systems that are based on polynomials or other functions as well as find roots and minimal values of functions.

5.7.1 Polynomials

The library functions for operating **polynomials** are declared in the header file `gsl_poly.h`. To find the roots of the polynomials, two methods are available: the analytical method for quadratic and cubic equations, and the iterative method for polynomials of the general type.

Consider the quadratic equation

$$x^2 - 2x + 1 = 0,$$

with one real root $x = 1$. The program to find the root is as follows:

Listing 5.66.

```
1  #include <stdio.h>
2  #include <gsl/gsl_poly.h>
3  int main (void){
4      int i;
5      // Define the  coefficients of the polynomial
6      double a = 1, b = -2, c = 1;
7      // Create variables and solve equation
8      double x0, x1;
9      i = gsl_poly_solve_quadratic(a, b, c, &x0, &x1);
10     if (i == 2) printf ("x0 = %+.10f\n x1 = %+.10f\n", x0, x1);
11     if (i == 1) printf ("x0 = %+.10f\n", x0);
12     if (i == 0) printf ("No solve");
13     return 0;
14 }
```

The function `gsl_poly_solve_quadratic(a, b, c, &x0, &x1)` takes as parameters the coefficients of the quadratic equation and variables to store the results. The function returns the number of real roots. In this case, the zero discriminant leads to two equal roots. If no real root is found, then the values of `x0`, `x1` are not modified. If only one real root exists, then its value is stored in `x0`, whereas `x1` is not modified.

The output of the program is

```
x0 = +1.0000000000
x1 = +1.0000000000
```

The library also has functions for the analytical solution of equations with complex roots and cubic equations.

To solve equations defined using the polynomial

$$P(x) = c_0 + c_1 x + c_2 x^2 + \cdots + c_n x^n = 0$$

the functions based on the iterative method are employed. As an example, we consider the equation $P(x) = x^4 - 1 = 0$ with the following program:

Listing 5.67.

```
1  #include <stdio.h>
2  #include <gsl/gsl_poly.h>
3  int main (void) {
4      // Define coefficients P(x) = -1 + x^4
5      double a[5] = {-1, 0, 0, 0, 1};
6      double z[8];
7      // Allocate memory
8      gsl_poly_complex_workspace *w = gsl_poly_complex_workspace_alloc(5);
9      // Solve
10     gsl_poly_complex_solve(a, 5, w, z);
11     // Free memory
12     gsl_poly_complex_workspace_free(w);
13     int i;
14     for (i = 0; i < 4; i++) {
15         printf("z%d = %+.10f %+.10f\n", i, z[2*i], z[2*i+1]);
16     }
17     return 0;
18 }
```

Here the coefficient of the highest order must be nonzero. To store the results, the array z is introduced. Its length is equal to twice the number of roots, since the roots are complex numbers. The function

Listing 5.68.

```
8  gsl_poly_complex_workspace *w = gsl_poly_complex_workspace_alloc(5);
```

allocates memory for a gsl_poly_complex_workspace structure. The size of this structure must be equal to the number of coefficients. The function

Listing 5.69.

```
12 gsl_poly_complex_workspace_free(w)
```

frees memory after solving the equation.

The function

Listing 5.70.

```
10  gsl_poly_complex_solve(a, 5, w, z)
```

finds the roots of the equation with a coefficient defined by the array a and returns the roots in the array z.

The output:

```
z0 = -1.0000000000 +0.0000000000
z1 = -0.0000000000 +1.0000000000
z2 = -0.0000000000 -1.0000000000
z3 = +1.0000000000 +0.0000000000
```

We can observe that four complex roots of the equation are found; the first number corresponds to the real part, the second one is the imaginary part.

5.7.2 One-dimensional root-finding

Here we discuss the functions for finding roots of arbitrary one- or multidimensional functions. These functions are declared within the header file gsl/gsl_roots.h.

The library contains iterative methods for finding **roots of one-dimensional functions**. Iterative methods are convenient for achieving the desired tolerance of the solution and viewing the intermediate iterations. There are two classes of finding root algorithms: the *root bracketing* and the *root polishing* methods.

Bracketing algorithms examine a bounded region where a root is located. To find the root with the desired tolerance, bracketing algorithms reduce the region in an iterative manner. This permits accurate estimation of an error of the solution. These algorithms can be applied if we have a single root or an odd number of roots in the region. However, only one root will be found.

Polishing algorithms attempt to improve an initial guess for the root at each iteration. Their convergence rate depends essentially on the initial guess for the root; they may diverge if this guess is not close enough. These algorithms have no restriction on the number of roots, but they also find only one root. Before using a polishing algorithm, we must visually analyze the function and indicate the correct initial guess.

To solve an equation using bracketing and polishing algorithms, we use the gsl_root_fsolver and gsl_root_fdfsolver structures, respectively. Both classes include three main steps during the iteration:
- initialization of solver state s for the algorithm T;
- updating the state s using the algorithm T;

– checking the state *s* for convergence and repeating the iteration if necessary.

Consider, for example, the quadratic equation

$$x^2 - 4 = 0,$$

and solve it using the Brent bracketing algorithm.

Listing 5.71.

```
1  #include <stdio.h>
2  #include <gsl/gsl_errno.h>
3  #include <gsl/gsl_math.h>
4  #include <gsl/gsl_roots.h>
5  // The equation parameters
6  struct quadratic_params {
7      double a, b, c;
8  };
9  // The equation definition
10 double quadratic(double x, void *params) {
11     struct quadratic_params *p = (struct quadratic_params*)params;
12     double a = p->a;
13     double b = p->b;
14     double c = p->c;
15     return (a * x + b) * x + c;
16 }
17 int main(void){
18     int status;
19     // Set the maximum number of iterations
20     int iter = 0, max_iter = 100;
21     // Set the initial guess of the root
22     // and the exact solution to test convergence
23     double r = 0, r_expected = 2.0;
24     // Set the region to find the root
25     double x_lo = 0.0, x_hi = 5.0;
26     // Set the equation coefficients
27     struct quadratic_params params = {1.0, 0.0, -4.0};
28     // Initialize the equation
29     gsl_function F;
30     F.function = &quadratic;
31     F.params = &params;
32     // Create the solver
33     const gsl_root_fsolver_type *T;
34     T = gsl_root_fsolver_brent;
35     // Define the state
36     gsl_root_fsolver *s;
37     s = gsl_root_fsolver_alloc (T);
38     // Define the solver parameters
39     gsl_root_fsolver_set (s, &F, x_lo, x_hi);
```

```
40    printf ("using %s method\n", gsl_root_fsolver_name (s));
41    printf ("%5s [%9s, %9s] %9s %10s %9s\n", "iter", "lower",
42          "upper", "root", "err", "err(est)");
43    do {
44        iter++;
45        // Perform the iteration
46        status = gsl_root_fsolver_iterate (s);
47        // Get the current value of the root
48        r = gsl_root_fsolver_root (s);
49        // Get the current bracketing interval for the solver
50        x_lo = gsl_root_fsolver_x_lower (s);
51        x_hi = gsl_root_fsolver_x_upper (s);
52        // Check for convergence
53        status = gsl_root_test_interval (x_lo, x_hi, 0, 0.001);
54        if (status == GSL_SUCCESS) printf ("Converged:\n");
55        printf ("%5d [%.7f, %.7f] %.7f %+.7f %.7f\n",iter, x_lo,
56                x_hi,r, r - r_expected,x_hi - x_lo);
57    } while (status == GSL_CONTINUE && iter < max_iter);
58    // Free memory
59    gsl_root_fsolver_free (s);
60    return status;
61 }
```

We start by preparing the function for `gsl_function`. For polishing algorithms, we must also specify the derivative of the function. Here we declare the type of the solver as the Brent bracketing algorithm:

Listing 5.72.

```
33 const gsl_root_fsolver_type *T;
34 T = gsl_root_fsolver_brent;
```

Table 5.10 presents the full list of available solvers.

Table 5.10. Root-finding algorithms.

Solver	Description
Bracketing algorithms	
gsl_root_fsolver_bisection	The bisection method
gsl_root_fsolver_falsepos	The false position method
gsl_root_fsolver_brent	The Brent method
Polishing algorithms	
gsl_root_fdfsolver_newton	The standard Newton method
gsl_root_fdfsolver_secants	The secant method (simplified Newton's method)
gsl_root_fdfsolver_steffenson	The Steffenson method

Using the function `gsl_root_fsolver_alloc`, we allocate memory for the state of the solver:

Listing 5.73.

```
36  gsl_root_fsolver *s;
37  s = gsl_root_fsolver_alloc (T);
```

The function `gsl_root_fsolver_set` specifies the parameters of the solver:

Listing 5.74.

```
39  gsl_root_fsolver_set(s, &F, x_lo, x_hi);
```

The function `gsl_root_fsolver_iterate(s)` performs the iteration and updates the state *s*. The function `gsl_root_test_interval` checks convergence and returns the status of the found root.

The output of the program:

```
using brent method
  iter [    lower,      upper]      root          err   err(est)
     1 [0.8000000, 5.0000000] 0.8000000  -1.2000000 4.2000000
     2 [0.8000000, 2.9000000] 2.9000000  +0.9000000 2.1000000
     3 [1.7081081, 2.9000000] 1.7081081  -0.2918919 1.1918919
     4 [1.7081081, 2.0546113] 2.0546113  +0.0546113 0.3465032
     5 [1.9957635, 2.0546113] 1.9957635  -0.0042365 0.0588477
     6 [1.9999429, 2.0546113] 1.9999429  -0.0000571 0.0546684
Converged:
     7 [1.9999429, 2.0000000] 2.0000000  +0.0000000 0.0000571
```

It is easy to see that the root of the equation is found with the given accuracy in 7 iterations.

5.7.3 Multidimensional root-finding

The **multidimensional root-finding** can be treated as solving a nonlinear system with *n* unknowns and *n* equations:

$$f_1(x_1, x_2, \ldots, x_n) = 0,$$
$$f_2(x_1, x_2, \ldots, x_n) = 0,$$
$$\ldots$$
$$f_n(x_1, x_2, \ldots, x_n) = 0.$$

In this case bracketing algorithms are not applied. All applicable algorithms are versions of the Newton method and use the initial guess. The structure of solvers permits combining them. The algorithms under onsideration are divided into two classes, i.e. the algorithms based on using derivatives, and the methods without employing derivatives.

For example, let us study the Rosenbrock system of equations:

$$f_1(x_1, x_2) = (1 - x_1),$$
$$f_2(x_1, x_2) = 100(x_2 - x_1^2),$$

which has one root at the point $(x_1, x_2) = (1, 1)$.

Listing 5.75.

```
1   #include <stdlib.h>
2   #include <stdio.h>
3   #include <gsl/gsl_vector.h>
4   #include <gsl/gsl_multiroots.h>
5   // Define the system of equations
6   struct rparams {
7       double a;
8       double b;
9   };
10  int rosenbrock_f(const gsl_vector *x, void *params, gsl_vector *f){
11      double a = ((struct rparams *) params)->a;
12      double b = ((struct rparams *) params)->b;
13      const double x0 = gsl_vector_get (x, 0);
14      const double x1 = gsl_vector_get (x, 1);
15      const double y0 = a * (1 - x0);
16      const double y1 = b * (x1 - x0 * x0);
17      gsl_vector_set (f, 0, y0);
18      gsl_vector_set (f, 1, y1);
19      return GSL_SUCCESS;
20  }
21  // The function for outputing the state of the solver
22  int print_state (size_t iter, gsl_multiroot_fsolver * s) {
23      printf ("iter = %3u x = % .3f % .3f "
24          "f(x) = % .3e % .3e\n",
25          iter,
26          gsl_vector_get (s->x, 0),
27          gsl_vector_get (s->x, 1),
28          gsl_vector_get (s->f, 0),
29          gsl_vector_get (s->f, 1));
30  }
31  int main (void) {
32      int status;
33      size_t i, iter = 0;
34      // Set the Rosenbrock system
```

```
35    const size_t n = 2;
36    struct rparams p = {1.0, 100.0};
37    gsl_multiroot_function f = {&rosenbrock_f, n, &p};
38    // Set the initial guess
39    double x_init[2] = {-10.0, -5.0};
40    gsl_vector *x = gsl_vector_alloc (n);
41    gsl_vector_set (x, 0, x_init[0]);
42    gsl_vector_set (x, 1, x_init[1]);
43    // Select the solver
44    const gsl_multiroot_fsolver_type *T = gsl_multiroot_fsolver_hybrids;
45    // Define the initial state
46    gsl_multiroot_fsolver *s = gsl_multiroot_fsolver_alloc (T, 2);
47    gsl_multiroot_fsolver_set (s, &f, x);
48    // Print the initial state
49    print_state (iter, s);
50    do {
51      iter++;
52      // Perform the iteration
53      status = gsl_multiroot_fsolver_iterate (s);
54      // Print the current state
55      print_state (iter, s);
56      // Check the solver status
57      if (status) break;
58      // Check the residual value
59      status = gsl_multiroot_test_residual (s->f, 1e-7);
60    } while (status == GSL_CONTINUE && iter < 1000);
61    printf ("status = %s\n", gsl_strerror (status));
62    // Free memory
63    gsl_multiroot_fsolver_free (s);
64    gsl_vector_free (x);
65    return 0;
66  }
```

First, we define the general system of equations with parameters:

Listing 5.76.

```
37  gsl_multiroot_function f = {&rosenbrock_f, n, &p};
```

For solvers which use the derivatives, we employ the structure `gsl_multiroot_function_fdf`. Next, we initialize the solver:

Listing 5.77.

```
44  const gsl_multiroot_fsolver_type* T = gsl_multiroot_fsolver_hybrids;
```

Table 5.11 demonstrates the list of available solvers.

Table 5.11. Multidimensional root-finding algorithms.

Solver	Description
Without derivatives	
gsl_multiroot_fsolver_hybrids	The Hybrid algorithm, which replaces calls to the Jacobian function by its finite difference approximation
gsl_multiroot_fsolver_hybrid	The Hybrid algorithm without internal scaling
gsl_multiroot_fsolver_dnewton	The discrete Newton algorithm
gsl_multiroot_fsolver_broyden	The Broyden algorithm
Using derivatives	
gsl_multiroot_fdfsolver_hybridsj	The Powell hybrid method
gsl_multiroot_fdfsolver_hybridj	The unscaled version of hybridsj
gsl_multiroot_fdfsolver_newton	The standard Newton method
gsl_multiroot_fdfsolver_gnewton	The modified Newton method

Here is the output of the program:

```
iter =    0 x = -10.000 -5.000 f(x) =  1.100e+01 -1.050e+04
iter =    1 x = -10.000 -5.000 f(x) =  1.100e+01 -1.050e+04
iter =    2 x = -3.976 24.826 f(x) =  4.976e+00  9.019e+02
iter =    3 x = -3.976 24.826 f(x) =  4.976e+00  9.019e+02
iter =    4 x = -3.976 24.826 f(x) =  4.976e+00  9.019e+02
iter =    5 x = -1.273 -5.681 f(x) =  2.273e+00 -7.302e+02
iter =    6 x = -1.273 -5.681 f(x) =  2.273e+00 -7.302e+02
iter =    7 x =  0.249  0.298 f(x) =  7.510e-01  2.358e+01
iter =    8 x =  0.249  0.298 f(x) =  7.510e-01  2.358e+01
iter =    9 x =  1.000  0.878 f(x) =  8.356e-11 -1.218e+01
iter =   10 x =  1.000  0.989 f(x) =  7.411e-12 -1.080e+00
iter =   11 x =  1.000  1.000 f(x) =  0.000e+00  0.000e+00
status = success
```

The error decreases with each iteration and the root is found in 11 iterations.

5.7.4 One-dimensional minimization

Minimization functions available in GSL as root-finding functions are iterative methods. Each algorithm stores its state, and therefore, we can switch between algorithms if necessary. Algorithms work in a bounded region and return only one minimum value at a time. The minimization functions are declared in the header file `gsl\gsl_min.h`.

The following is an example of finding the minimal value of the function $f(x) = \cos(x) - 1$ is presented:

Listing 5.78.

```
1  #include <stdio.h>
2  #include <gsl/gsl_errno.h>
3  #include <gsl/gsl_math.h>
4  #include <gsl/gsl_min.h>
5  // Define function
6  double f(double x, void * params) {
7      return cos(x) - 1.0;
8  }
9  int main (void) {
10     int status;
11     // Set the maximum number of iterations
12     int iter = 0, max_iter = 100;
13     // Set the exact solution for the convergence test
14     double m = 2.0, m_expected =  M_PI;
15     // Set the region
16     double a = 0.0, b = 6.0;
17     // Create the function
18     gsl_function F;
19     F.function = &f;
20     F.params = 0;
21     // Set a minimization type
22     const gsl_min_fminimizer_type *T;
23     T = gsl_min_fminimizer_brent;
24     // Initialize the state of the solver
25     gsl_min_fminimizer *s;
26     s = gsl_min_fminimizer_alloc (T);
27     gsl_min_fminimizer_set (s, &F, m, a, b);
28     printf ("using %s method\n", gsl_min_fminimizer_name (s));
29     printf ("%5s [%9s, %9s] %9s %10s %9s\n", "iter",
30         "lower", "upper", "min", "err", "err(est)");
31     printf ("%5d [%.7f, %.7f] %.7f %+.7f %.7f\n", iter, a, b,
32         m, m - m_expected, b - a);
33     do {
34         iter++;
35         // Perform the iteration
36         status = gsl_min_fminimizer_iterate (s);
37         // Get the current value of the minimum
38         // and the current lower and upper bounds
39         m = gsl_min_fminimizer_x_minimum (s);
40         a = gsl_min_fminimizer_x_lower (s);
41         b = gsl_min_fminimizer_x_upper (s);
```

```
42        // Get the state status
43        status = gsl_min_test_interval (a, b, 0.001, 0.0);
44        if (status == GSL_SUCCESS)
45        printf ("Converged:\n");
46        printf ("%5d [%.7f, %.7f] %.7f %+.7f %.7f\n", iter, a, b,
47            m, m - m_expected, b - a);
48    } while (status == GSL_CONTINUE && iter < max_iter);
49    // Free memory
50    gsl_min_fminimizer_free (s);
51    return status;
52 }
```

We begin by selecting the minimization algorithm:

Listing 5.79.

```
22 const gsl_min_fminimizer_type *T;
23 T = gsl_min_fminimizer_brent;
```

Table 5.12 presents the list of available algorithms. Next, we allocate memory for the selected minimizer and specify its parameters:

Listing 5.80.

```
25 gsl_min_fminimizer *s;
26 s = gsl_min_fminimizer_alloc (T);
27 gsl_min_fminimizer_set (s, &F, m, a, b);
```

Further, we perform the iteration using the function `gsl_min_fminimizer_iterate` and get the current value of the minimum and the current lower and upper bounds of the interval.

Listing 5.81.

```
39 m = gsl_min_fminimizer_x_minimum(s);
40 a = gsl_min_fminimizer_x_lower(s);
41 b = gsl_min_fminimizer_x_upper(s);
```

Table 5.12. One-dimensional minimization algorithms.

Algorithm	Description
gsl_min_fminimizer_goldensection	The golden section algorithm
gsl_min_fminimizer_brent	The Brent minimization algorithm, which combines parabolic interpolation with the golden section algorithm
gsl_min_fminimizer_quad_golden	The Brent algorithm, whichuses the safeguarded step-length algorithm of Gill and Murray.

The output of the program is

```
using brent method
 iter [    lower,      upper]       min          err    err(est)
    0 [0.0000000, 6.0000000] 1.0000000 -2.1415927 6.0000000
    1 [1.0000000, 6.0000000] 2.9098300 -0.2317627 5.0000000
    2 [1.0000000, 3.7455735] 2.9098300 -0.2317627 2.7455735
    3 [2.9098300, 3.7455735] 3.0739632 -0.0676294 0.8357435
    4 [3.0739632, 3.7455735] 3.1441015 +0.0025089 0.6716103
    5 [3.0739632, 3.1441015] 3.1417797 +0.0001870 0.0701383
    6 [3.0739632, 3.1417797] 3.1415922 -0.0000005 0.0678164
Converged:
    7 [3.1415922, 3.1417797] 3.1415927 -0.0000000 0.0001875
```

The minimal value with the given accuracy 0.001 is obtained in 7 iterations.

5.7.5 Multidimensional minimization

Now we consider the functions of finding the **minimal values of multidimensional functions**. The GSL library provides iterative methods of minimization. Minimizers can be combined, since the intermediate states between iterations are fully stored. Minimization algorithms can find only one value of the minimum. If there is more than one value, then a minimizer returns the first value.

In this case, only polishing algorithms are used to find the minimal value. The polishing algorithms are divided into two classes: with and without using derivatives.

As an example, we consider the function

$$f(x, y) = \frac{x^2}{4} + \frac{y^2}{4},$$

with the minimal value 0 at the point $(0, 0)$.

Listing 5.82.

```c
#include <stdlib.h>
#include <stdio.h>
#include <gsl/gsl_vector.h>
#include <gsl/gsl_multimin.h>
// Define function
double my_f(const gsl_vector *v, void *params) {
    double x, y;
    x = gsl_vector_get(v, 0);
    y = gsl_vector_get(v, 1);
    double *p = (double *)params;
    return p[2] * (x - p[0]) * (x - p[0]) + p[3] * (y - p[1]) *
           (y - p[1]) + p[4];
}
void my_df(const gsl_vector *v, void *params, gsl_vector *df) {
    double x, y;
    double *p = (double *)params;
    x = gsl_vector_get(v, 0);
    y = gsl_vector_get(v, 1);
    gsl_vector_set(df, 0, 2.0 * p[2] * (x - p[0]));
    gsl_vector_set(df, 1, 2.0 * p[3] * (y - p[1]));
}
void my_fdf (const gsl_vector *x, void *params, double *f,
             gsl_vector *df) {
    *f = my_f(x, params);
    my_df(x, params, df);
}
int main(void) {
    size_t iter = 0;
    int status;
    // Set the function coefficients
    // [2]*(x - p[0])^2 + p[3] * (y - p[1]) + p[4]
    double par[5] = { 0.0, 0.0, 0.25, 0.25, 0.0 };
    // Define the function
    gsl_multimin_function_fdf my_func;
    my_func.n = 2;
    my_func.f = my_f;
    my_func.df = my_df;
    my_func.fdf = my_fdf;
    my_func.params = par;
    // Set the initial guess
    gsl_vector *x = gsl_vector_alloc (2);
    gsl_vector_set (x, 0, 5.0);
    gsl_vector_set (x, 1, 7.0);
    // Choose the solver
    const gsl_multimin_fdfminimizer_type *T =
        gsl_multimin_fdfminimizer_conjugate_fr;
    // Define the initial state
```

```
48     gsl_multimin_fdfminimizer *s =
49         gsl_multimin_fdfminimizer_alloc (T, 2);
50     gsl_multimin_fdfminimizer_set (s, &my_func, x, 0.01, 1e-4);
51     do {
52         iter++;
53         // Perform the iteration
54         status = gsl_multimin_fdfminimizer_iterate (s);
55         if (status) break;
56         status = gsl_multimin_test_gradient (s->gradient, 1e-3);
57         if (status == GSL_SUCCESS) printf ("Minimum found at:\n");
58         // Print the current state
59         printf ("%5d %.5f %.5f %10.5f\n", iter,
60             gsl_vector_get (s->x, 0),
61             gsl_vector_get (s->x, 1),
62             s->f);
63     } while (status == GSL_CONTINUE && iter < 100);
64     // Free memory
65     gsl_multimin_fdfminimizer_free (s);
66     gsl_vector_free (x);
67     return 0;
68 }
```

First, we describe the structure my_fdf for the function and its derivative. Further, we define the function and specify the parameters.

After setting the initial guess, we select the minimization algorithm (the available algorithms are listed in Table 5.13):

Listing 5.83.

```
45 const gsl_multimin_fdfminimizer_type *T =
46     gsl_multimin_fdfminimizer_conjugate_fr;
```

Next, we allocate memory for the selected minimizer and set its parameters:

Listing 5.84.

```
48 gsl_multimin_fdfminimizer *s =
49     gsl_multimin_fdfminimizer_alloc (T, 2);
50 gsl_multimin_fdfminimizer_set (s, &my_func, x, 0.01, 1e-4);
```

Then we run the loop until the minimal value is found. Inside the loop, we perform the iteration gsl_multimin_fdfminimizer_iterate and check the state of the minimizer.

Table 5.13. Multidimensional minimization algorithms.

Algorithm	Description
With derivatives	
gsl_multimin_fdfminimizer_conjugate_fr	The Fletcher-Reeves conjugate gradient algorithm
gsl_multimin_fdfminimizer_conjugate_pr	The Polak-Ribiere conjugate gradient algorithm
gsl_multimin_fdfminimizer_vector_bfgs	The vector Broyden-Fletcher-Goldfarb-Shanno (BFGS) algorithm
gsl_multimin_fdfminimizer_vector_bfgs2	The modified vector Broyden-Fletcher-Goldfarb-Shanno (BFGS) algorithm
gsl_multimin_fdfminimizer_steepest_descent	The steepest descent algorithm
Without derivatives	
gsl_multimin_fminimizer_nmsimplex	The simplex algorithm of Nelder and Mead
gsl_multimin_fminimizer_nmsimplex2	The modified simplex algorithm of Nelder and Mead
gsl_multimin_fminimizer_nmsimplex2rand	The variation of the nmsimplex2

The following is the output of the program:

```
   1 4.99419 6.99186    18.45701
   2 4.98256 6.97559    18.37119
   3 4.95931 6.94304    18.20014
   4 4.91281 6.87794    17.86045
   5 4.81982 6.74774    17.19066
   6 4.63382 6.48735    15.88949
   7 4.26183 5.96656    13.44075
   8 3.51784 4.92498     9.15766
   9 2.02987 2.84182     3.04908
  10 -0.94607 -1.32449    0.66233
Minimum found at:
  11 0.00000 -0.00000     0.00000
```

The minimal value of the function is achieved in 11 iterations.

5.8 Interpolation and approximation of functions

This section describes the functions and methods for approximate reconstructing functions. Namely, we consider interpolation, least-squares fitting, and basis splines.

5.8.1 Interpolation

The **interpolation** is a method of constructing a function $f(x)$ using the known values at several points of an interval. The interpolation functions are declared in the header files $gsl\_interp.h$ and $gsl\_spline.h$.

We now consider an example with interpolation of data using the cubic spline, where 5 points are given: $x_i = i$, $y_i = i + e^i$, $i = 0, 1, 2, 3, 4$:

Listing 5.85.

```
1  #include <stdlib.h>
2  #include <stdio.h>
3  #include <math.h>
4  #include <gsl/gsl_errno.h>
5  #include <gsl/gsl_interp.h>
6  #include <gsl/gsl_spline.h>
7  int main ( void ) {
8      int i, N=5;
9      //Define arrays to store input data
10     double x[N], y[N];
11     //Define the variables for the unknowns
12     double xi, yi;
13     //Set input data
14     printf("Input data\n");
15     for (i = 0; i < N; i++) {
16         x[i] = i;
17         y[i] = i + exp(i);
18         printf("%g %g\n", x[i], y[i]);
19     }
20     // Set interpolation type
21     const gsl_interp_type *t = gsl_interp_linear;
22     // Create the object interp with type t and size N
23     gsl_interp *interp = gsl_interp_alloc(t, N);
24     // Initialize the object with input data x, y
25     gsl_interp_init(interp, x, y, N);
26     // Create the accelerator
27     gsl_interp_accel *acc = gsl_interp_accel_alloc();
28     printf("Interpolated data\n");
29     for (xi = x[0]; xi < x[N-1]; xi += 0.5) {
30         // Calculate interpolated value  yi for point xi
31         yi = gsl_interp_eval(interp, x, y, xi, acc);
```

```
32          // Print result
33          printf("%g %g\n", xi, yi);
34      }
35      //Free memory
36      gsl_interp_free(interp);
37      gsl_interp_accel_free(acc);
38      return 0;
39  }
```

For each value $x_{i+1/2} = x_i + 0.5$, we get the following values of y_i:

```
Input data
0 1
1 3.71828
2 9.38906
3 23.0855
4 58.5982
Interpolated data
0 1
0.5 2.35914
1 3.71828
1.5 6.55367
2 9.38906
2.5 16.2373
3 23.0855
3.5 40.8418
```

The interpolation function for a given data set is stored in the object `gsl_interp`. To create new interpolation functions, we apply the function

Listing 5.86.

```
23  gsl_interp *interp = gsl_interp_alloc(t, N);
```

which allocates memory for interpolating object of type `t` with `N` given points. Next, to initialize the object `interp`, we call

Listing 5.87.

```
25  gsl_interp_init(interp, x, y, N);
```

At the end, the object `interp` must be removed:

Listing 5.88.

```
36 gsl_interp_free(interp);
```

In this example, we use the cubic spline interpolation method:

Listing 5.89.

```
21 const gsl_interp_type *t = gsl_interp_cspline;
```

GSL provides the standard interpolation types, such as linear interpolation, polynomial interpolation, cubic spline interpolation, Akima spline interpolation and so on. These interpolation types are interchangeable. Interpolation can be defined for both natural and periodic boundary conditions. Table 5.14 presents the list of available interpolation types.

Table 5.14. Interpolation types.

Type	Description
gsl_interp_linear	Linear interpolation
gsl_interp_polynomial	Polynomial interpolation
gsl_interp_cspline	Cubic spline with natural boundary conditions
gsl_interp_cspline_periodic	Cubic spline with periodic boundary conditions
gsl_interp_akima	Nonrounded Akima spline with natural boundary conditions
gsl_interp_akima_periodic	Nonrounded Akima spline with periodic boundary conditions

The state of an interpolation process is stored in an object `gsl_interp_accel`, which is a sort of iterator for interpolation. It stores the previous state of the process. Thus, if the subsequent interpolation point lies in the same range, then its index value can be returned instantly.

To create an accelerator object, it is necessary to apply the function

Listing 5.90.

```
27 gsl_interp_accel *acc = gsl_interp_accel_alloc();
```

This object must also be removed:

Listing 5.91.

```
37  gsl_interp_accel_free(acc);
```

To obtain the interpolated value y_i, we call the function

Listing 5.92.

```
31  yi = gsl_interp_eval(interp, x, y, xi, acc);
```

Here `interp` is the interpolation object, `x,y` denotes a given data set, `xi` is a new value of `x`, and `acc` is the accelerator.

There are also additional functions for calculating derivatives and integrals of interpolated functions (see Table 5.15).

Table 5.15. Functions for calculating derivatives and integrals.

Function	Description
gsl_interp_eval_deriv	The derivative of an interpolated function for a given point x
gsl_interp_eval_deriv2	The second derivative of an interpolated function for a given point x
gsl_interp_eval_integ	The numerical integral of an interpolated function over the range (a, b)

These functions require pointers to arrays `x` and `y` at each call. As a simplification, `GSL` provides functions that are equivalent to the corresponding functions of `gsl_interp` and store a copy of the data in the object `gsl_spline`. This eliminates the need to enter values of `x` and `y` as arguments for each calculation. These functions are defined in the header file `gsl_spline.h`.

5.8.2 Least-squares method

The interpolation of large experimental data obtained with some errors is irrational. The **least squares method** is an approach that determines an approximate function by minimizing the sum of the squared deviations of the values of this function and the experimental data.

Functions of the least-squares method are declared in the header file `gsl_fit.h`.

The approximate function $Y(c, x)$ is found by minimizing χ^2, i.e. using the weighted sum of squared residuals over n given points (x_i, y_i) for the function $Y(c, x)$:

$$\chi^2 = \sum_i w_i \left(y_i - Y(c_i, w_i) \right)^2 .$$

Here $w_i = 1/\sigma_i^2$ is a weight factor, and σ_i stands for the experimental error y_i. For unweighted data, the sum χ^2 is calculated without any weight factors.

The fitting functions return the best parameters $c = c_0, c_1, \ldots$ and corresponding covariance matrix, which measures the statistical errors.

The following is the program that calculates the linear interpolated function $Y = Y(c, x) = c_0 + c_1 x$ for the given data set using the least-square method:

Listing 5.93.

```
1  #include <stdio.h>
2  #include <gsl/gsl_fit.h>
3  int main(void) {
4      int i, n = 4;
5      //Define input data
6      double x[4] = { 1, 2, 3, 4 };
7      double y[4] = { 11, 13, 12, 14 };
8      double w[4] = { 0.1, 0.2, 0.3, 0.4 };
9      // Declare variables for parameters of the approximate function
10     double c0, c1, cov00, cov01, cov11, chisq;
11     // Calculate the coefficients of the linear regression function
12     // and covariance matrix for the weighted data (x,y)
13     gsl_fit_wlinear(x, 1, w, 1, y, 1, n,
14         &c0, &c1, &cov00, &cov01, &cov11, &chisq);
15     // Print the input data and weight coefficients
16     for (i = 0; i < n; i++)
17         printf("data: %g %g %g\n", x[i], y[i], 1/sqrt(w[i]));
18     // Print the best fit function
19     printf("best fit: ");
20     printf("Y = %g + %g X\n", c0, c1);
21     // Print the covariance matrix and the sum of squared residuals
22     printf("covariance matrix:\n");
23     printf("[ %g, %g\n %g, %g]\n", cov00, cov01, cov01, cov11);
24     printf("chisq = %g\n", chisq);
25     return 0;
26  }
```

The output is as follows:

```
data: 1 11 3.16228
data: 2 13 2.23607
data: 3 12 1.82574
data: 4 14 1.58114
best fit: Y = 10.5 + 0.8 X
covariance matrix:
[ 10, -3
  -3,  1]
chisq = 0.45
```

In this example we use the function which evaluates the coefficient of the linear regression $Y(c,x) = c_0 + c_1 x$ for the weighted data:

Listing 5.94.

```
13  gsl_fit_wlinear(x, 1, w, 1, y, 1, n,
14       &c0, &c1, &cov00,   &cov01, &cov11, &chisq);
```

Here x and y is the set of weighted data of length n, 1 is a stride, w is the weight function in the form of the vector of length n with the stride = 1. The covariance matrix (c0, c1) is estimated from the scattering of points around the function and returned via the parameters (cov00, cov01, cov11), whereas chisq is the weighted sum of squared residuals χ^2.

If we want to calculate the coefficients c_0, c_1 of the best approximation $Y(c,x) = c_0 + c_1 x$, for data without weight, we use the function gsl_fit_linear.

Using the coefficients (c0, c1) and the covariance parameters (cov00, cov01, cov11), we can calculate the value y of the linear regression $Y(c,x) = c_0 + c_1 x$ at the point x and the standard deviation using the function gsl_fit_linear_est. Similarly, there are functions to determine the coefficients of the function $Y = c_1 X$.

The library provides functions for least-square multiparameters fitting using the function of type $Y = Xc$, where Y is a vector of n data sets, X is an $n \times p$ matrix of parameter,s and c is the vector of best-fit parameters. The value χ^2 is calculated as

$$\chi^2 = \sum_i w_i (y_i - \sum_j X_{ij} c_j)^2.$$

To apply this fitting, we need to formulate the $n \times p$ matrix X defined by any number of functions and variables. For example, to fit to the polynomial x with p-th order, the matrix is defined as

$$X_{ij} = x_i^j, \quad i = 0, 1, \ldots, n, \quad j = 0, 1, \ldots, p - 1.$$

To fit to a set of sinusoidal functions with fixed frequencies $w_1, w_2, \ldots, w_p$, we apply

$$X_{ij} = \sin(w_j x_i), \quad i = 0, 1, \ldots, n, \quad j = 0, 1, \ldots, p.$$

To fit to independent variables $x_1, x_2, \ldots, x_p$, we use

$$X_{ij} = x_j(i),$$

where $x_j(i)$ is the i-th value of the given function.

To apply these functions, we must include the header file `gsl_multifit.h`.

5.8.3 B-splines

B-splines are used as basis functions to build smoothing curves for a large set of data. B-splines differ from interpolation splines in that the resulting curve is not required to pass through given points.

To build a B-spline, the abscissa axis is divided into some number of intervals, where the endpoints of each interval are called breakpoints. These points are then converted to knots by setting various continuity and smoothness conditions at each interface.

Basis splines of order k are defined by

$$B_{i,1}(x) = \begin{cases} 1, & x_i \leq x < x_{i+1}, \\ 0, & \text{otherwise,} \end{cases} \tag{5.1}$$

$$B_{i,k}(x) = \frac{(x - x_i)}{x_{i+k-1} - x_i} B_{i,k-1}(x) + \frac{(x_{i+k} - x)}{(x_{i+k} - x_{i+1})} B_{i+1,k-1}(x) \tag{5.2}$$

for $i = 0, 1, \ldots, n - 1$. For $k = 4$, we get a cubic B-spline. This recurrence relation can be numerically evaluated using the de Boor algorithm.

If appropriate knots are defined on an interval $[a, b]$, then the basis functions of the B-spline generate a complete set on this interval. Therefore, we can extend smoothing functions as

$$f(x) = \sum_{i=0}^{n-1} c_i B_{i,k}(x)$$

by setting the data pairs $(x_j, f(x_j))$. The coefficients c_i can be easily obtained using the least-squares method.

Let us employ the linear least-square approximation based on the basis functions of cubic B-splines with uniform breakpoints. The data is generated by the function with Gaussian noise:

$$y(x) = \cos(x) \quad x \in [0, 10].$$

Listing 5.95.

```
1   #include <stdio.h>
2   #include <stdlib.h>
3   #include <math.h>
4   #include <gsl/gsl_bspline.h>
5   #include <gsl/gsl_multifit.h>
6   #include <gsl/gsl_rng.h>
7   #include <gsl/gsl_randist.h>
8   #include <gsl/gsl_statistics.h>
9   // Number of data points
10  #define N 100
11  // Number of coefficients
12  #define NCOEFFS 12
13  // Breakpoints nbreak = ncoeffs + 2 - k when k = 4
14  #define NBREAK (NCOEFFS - 2)
15  int main (void) {
16      const size_t n = N;
17      const size_t ncoeffs = NCOEFFS;
18      const size_t nbreak = NBREAK; size_t i, j;
19      double dy;
20      gsl_rng *r;
21      gsl_vector *c, *w;
22      gsl_vector *x, *y;
23      gsl_matrix *X, *cov;
24      gsl_bspline_workspace *bw;
25      gsl_vector *B;
26      gsl_multifit_linear_workspace *mw;
27      double chisq, Rsq, dof, tss;
28      gsl_rng_env_setup();
29      r = gsl_rng_alloc(gsl_rng_default);
30      // Allocate memory for B-spline (k = 4)
31      bw = gsl_bspline_alloc(4, nbreak);
32      B = gsl_vector_alloc(ncoeffs);
33      x = gsl_vector_alloc(n);
34      y = gsl_vector_alloc(n);
35      X = gsl_matrix_alloc(n, ncoeffs);
36      c = gsl_vector_alloc(ncoeffs);
37      w = gsl_vector_alloc(n);
38      cov = gsl_matrix_alloc(ncoeffs, ncoeffs);
39      mw = gsl_multifit_linear_alloc(n, ncoeffs);
40      // Set input data
41      for (i = 0; i < n; ++i){
42          double sigma;
43          double xi = (10.0 / (N - 1)) * i;
44          double yi = cos(xi);
45          sigma = 0.1 * yi;
46          dy = gsl_ran_gaussian(r, sigma);
47          yi += dy;
```

```
48        gsl_vector_set(x, i, xi);
49        gsl_vector_set(y, i, yi);
50        gsl_vector_set(w, i, 1.0 / (sigma * sigma));
51        printf("%f %f\n", xi, yi);
52    }
53    // Set uniform breakpoints on [0, 10]
54    gsl_bspline_knots_uniform(0.0, 10.0, bw);
55    // Construct the matrix  X
56    for (i = 0; i < n; ++i){
57        double xi = gsl_vector_get(x, i);
58        // Compute B_j(xi)  for all j
59        gsl_bspline_eval(xi, B, bw);
60        // Fill in  i-th row of  matrix X
61        for (j = 0; j < ncoeffs; ++j){
62            double Bj = gsl_vector_get(B, j);
63            gsl_matrix_set(X, i, j, Bj);
64        }
65    }
66    // Calculate approximation
67    gsl_multifit_wlinear(X, w, y, c, cov, &chisq, mw);
68    dof = n - ncoeffs;
69    tss = gsl_stats_wtss(w->data, 1, y->data, 1, y->size);
70    Rsq = 1.0 - chisq / tss;
71    fprintf(stderr, "chisq/dof = %e,Rsq = %f\n", chisq / dof, Rsq);
72    // Construct the smoothing curve
73    double xi, yi, yerr;
74    for (xi = 0.0; xi < 10.0; xi += 0.1){
75        gsl_bspline_eval(xi, B, bw);
76        gsl_multifit_linear_est(B, c, cov, &yi, &yerr);
77        printf("%f %f\n", xi, yi);
78    }
79    // Free memory
80    gsl_rng_free(r);
81    gsl_bspline_free(bw);
82    gsl_vector_free(B);
83    gsl_vector_free(x);
84    gsl_vector_free(y);
85    gsl_matrix_free(X);
86    gsl_vector_free(c);
87    gsl_vector_free(w);
88    gsl_matrix_free(cov);
89    gsl_multifit_linear_free(mw);
90    return 0;
91 }
```

The solution of this problem is shown in Figure 5.1.

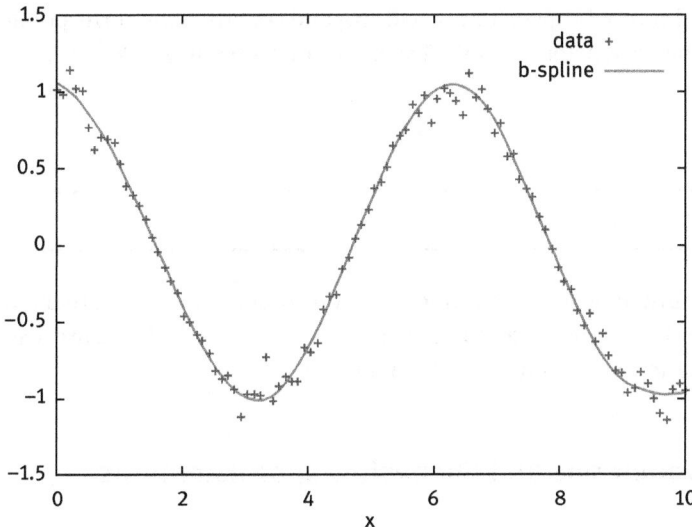

Fig. 5.1. The result of constructing cubic B-spline.

To evaluate the basis functions of the B-spline, we begin by allocating memory:

Listing 5.96.

```
24  gsl_bspline_workspace *bw;
25  . . .
26  bw = gsl_bspline_alloc(4, nbreak);
```

At the end, we must free memory by calling the function

Listing 5.97.

```
81  gsl_bspline_free(bw);
```

If we need to use derivatives of the B-spline, we can also employ the function gsl_bspline_deriv_workspace to allocate memory and the function gsl_bspline_deriv_free to free memory.

Using the function

Listing 5.98.

```
54  gsl_bspline_knots_uniform(0.0, 10.0, bw);
```

we set uniform breakpoints on the given interval [0, 10] and construct the corresponding knot vector using the parameter `nbreak`. The knots are stored in `w->knots`.

The function

Listing 5.99.

```
75  gsl_bspline_eval(xi, B, bw);
```

evaluates all the basis functions of the B-spline at the position `xi` and stores them in the vector `B`. The vector `B` must be of length `n = nbreak + k - 2`. There are also functions for the evaluation of basis functions derivatives.

5.9 Fast Fourier transforms and Chebyshev approximations

The `GSL` library provides functions for performing fast Fourier transforms and Chebyshev approximations.

5.9.1 Fast Fourier transform

Fast Fourier transforms are algorithms for calculating the discrete Fourier transform

$$x_j = \sum_{k=0}^{n-1} z_k \exp(-2\pi ijk/n).$$

Discrete Fourier transforms usually arise when approximating the continuous Fourier transforms, where a function is given at discrete intervals. The discrete Fourier transform can be presented as a matrix-vector multiplication Wz. For a vector of length n, this multiplication takes $O(n^2)$ operations. A fast Fourier transform factorizes the matrix W into smaller submatrices. If n can be presented as a product $n_1 n_2 \cdots n_m$, then a discrete Fourier transform can be calculated in $O(n \sum_i n_i)$ operations. If $n_i = 2$, then the discrete fast Fourier transform is performed in $O(n \log_2 n)$ operations.

All fast Fourier transform functions are divided into three types: forwards, inverse, and backwards. The definition of the forward Fourier transform is

$$x_j = \sum_{k=0}^{n-1} z_k \exp(-2\pi ijk/n),$$

and the definition of the inverse Fourier transform is

$$z_j = \frac{1}{n} \sum_{k=0}^{n-1} x_k \exp(2\pi ijk/n).$$

Thus, a call to the function `gsl_fft_complex_forward`, followed by a call to the function `gsl_fft_complex_inverse`, must return the input data with numerical errors.

The backward transform corresponds to the inverse transform without the factor $1/n$; it is used to avoid additional division if there are no restrictions on the result of the function.

The library supports both radix-2 and mixed-radix fast Fourier transform functions.

Here we will consider fast Fourier transform realization for a pulse. For this example, we apply the fast Fourier transform function for complex data. This transform uses the radix-2 Cooley–Tukey algorithm.

Listing 5.100.

```c
#include <stdio.h>
#include <math.h>
#include <gsl/gsl_fft_complex.h>
// Define real and imagine parts
#define Re(z, i) ((z)[2 * (i)])
#define Im(z, i) ((z)[2 * (i) + 1])
int main() {
    // Set pulse parameters: interval and length
    int Q = 128;
    int R = 108;
    // Declare data array
    double data[2 * Q];
    // Data initialization
    for (int i = 0; i < Q; i++) {
        // Set input data
        Re(data, i) = 0.0;
        Im(data, i) = 0.0;
        if (fabs(Q / 2 - i) < R / 2)
            Re(data, i) = 1.0;
        // Print input data
        printf("%d %e %e\n", i, Re(data, i), Im(data, i));
    }
    printf("\n");
    // Forward Fast Fourier transform
    gsl_fft_complex_radix2_forward(data, 1, Q);
    // Print transformed data
    for (int i = 0; i < Q; i++) {
        printf("%d %e %e\n", i, Re(data, i) / sqrt(Q),
            Im(data, i) / sqrt(Q));
    }
    printf("\n");
    // Inverse Fast Fourier transform
    gsl_fft_complex_radix2_inverse(data, 1, Q);
```

```
34      // Print data after inverse transform
35      for (int i = 0; i < Q; i++) {
36          printf("%d %e %e\n", i, Re(data, i), Im(data, i));
37      }
38      return 0;
39  }
```

To employ functions of fast Fourier transforms, we include the header file:

Listing 5.101.

```
3  #include <gsl/gsl_fft_complex.h>
```

Next, to perform a fast Fourier transform, we use the following function:

Listing 5.102.

```
25  gsl_fft_complex_radix2_forward(data, 1, Q);
```

Here `data` is data, 1 is a stride, and `Q` is the length of the transform. The stride defines the efficiency of data processing. For example, if the stride value equals 2, then each second value will be processed.

For the inverse fast Fourier transform, we apply a similar function:

Listing 5.103.

```
33  gsl_fft_complex_radix2_inverse(data, 1, Q);
```

The result is depicted in Figure 5.2.

The library also provides a function for performing Fast Fourier transforms for real data.

5.9.2 Chebyshev approximations

A **Chebyshev approximation** is a truncation of the series

$$f(x) = \sum_{n=1}^{\infty} c_n T_n(x),$$

where the Chebyshev polynomials

$$T_n(x) = \cos(n \arccos x)$$

form an orthogonal basis of polynomials on the interval $[-1, 1]$.

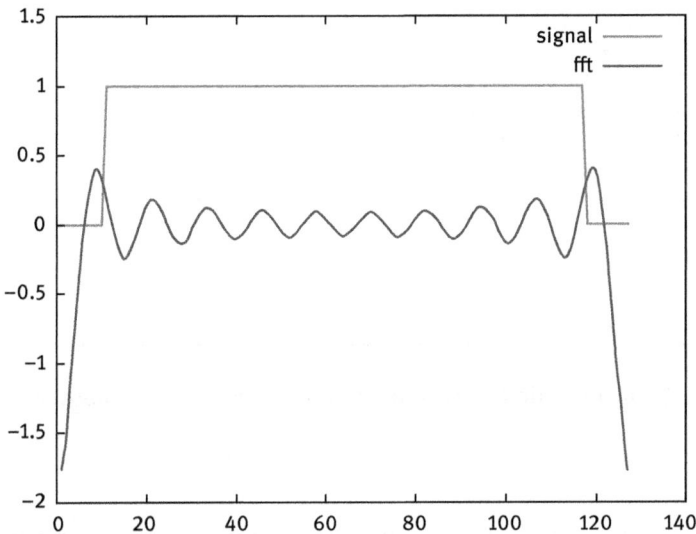

Fig. 5.2. The pulse and its fast Fourier transform.

Let us now consider how to apply Chebyshev approximations.

Listing 5.104.

```
1  #include <stdio.h>
2  #include <gsl/gsl_math.h>
3  #include <gsl/gsl_chebyshev.h>
4  // The function
5  double f(double x, void *p) {
6      if (x < 0.25)
7          return 0.0;
8      else if (x < 0.75)
9          return 1.0;
10     else
11         return 0.0;
12 }
13 int main(void) {
14     // Set the number of points for polynomial
15     int i, n = 10000;
16     // Allocate memory
17     gsl_cheb_series *cs = gsl_cheb_alloc(40);
18     gsl_function F;
19     F.function = f;
20     F.params = 0;
21     // Calculate the Chebyshev approximation
22     // for the function over the range [0,1]
23     gsl_cheb_init(cs, &F, 0.0, 1.0);
```

```
24      for (i = 0; i < n; i++) {
25          double x = 1.0 * i / n;
26          // Evaluate the Chebyshev series
27          double r10 = gsl_cheb_eval_n(cs, 10, x);
28          double r40 = gsl_cheb_eval(cs, x);
29          printf("%g %g %g %g\n", x, GSL_FN_EVAL(&F, x), r10, r40);
30      }
31      // Free memory
32      gsl_cheb_free(cs);
33      return 0;
34  }
```

First, we must include the header file for working with the functions of Chebyshev approximations:

Listing 5.105.

```
3  #include <gsl/gsl_chebyshev.h>
```

Now we set the function similar to the pulse function from the example of the fast Fourier transform:

Listing 5.106.

```
5  double f(double x, void *p) {
6      if (x < 0.25)
7          return 0.0;
8      else if (x < 0.75)
9          return 1.0;
10     else
11         return 0.0;
12 }
```

Next, we need to allocate memory for the Chebyshev series, for example of the order 40:

Listing 5.107.

```
17 gsl_cheb_series *cs = gsl_cheb_alloc(40);
```

Since Chebyshev polynomials work with special functions, we link the given function with the necessary GSL function:

Listing 5.108.

```
18  gsl_function F;
19  F.function = f;
20  F.params = 0;
```

We calculate the Chebyshev approximation for the function over the range [0, 1]. To do this, we use the function `gsl_cheb_init`. The function takes a pointer to the Chebyshev series, the function f, and the range:

Listing 5.109.

```
23  gsl_cheb_init(cs, &F, 0.0, 1.0);
```

Further, we evaluate the Chebyshev series at each point:

Listing 5.110.

```
24  for (i = 0; i < n; i++) {
25      double x = 1.0 * i / n;
26      . . .
27      double r10 = gsl_cheb_eval_n(cs, 10, x);
28      double r40 = gsl_cheb_eval(cs, x);
```

At the end, we free allocated memory:

Listing 5.111.

```
32  gsl_cheb_free(cs);
```

The result is presented in Figure 5.3.

The main functions for working with Chebyshev polynomials are presented in Table 5.16.

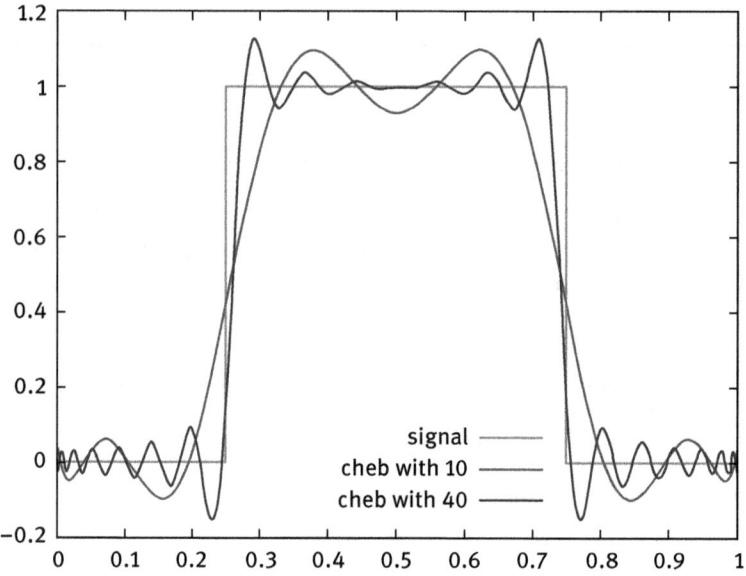

Fig. 5.3. Input function and its Chebyshev approximation.

Table 5.16. Functions for Chebyshev approximations.

Function	Description
gsl_cheb_order	The function returns the order of Chebyshev series
gsl_cheb_size	The function returns the size of the Chebyshev coefficient array
gsl_cheb_eval	The function evaluates the Chebyshev series at a given point
gsl_cheb_eval_err	The function evaluates the Chebyshev series at a given point and returns the estimates for the result and its absolute error
gsl_cheb_eval_n	The function returns the Chebyshev series with the given order
gsl_cheb_eval_n_err	Similar to gsl_cheb_eval_n
gsl_cheb_calc_deriv	The function calculates the derivative of the Chebyshev series
gsl_cheb_calc_integ	The function calculates the integral of the Chebyshev series

Ivan K. Sirditov

6 Visualization of computed data

Abstract: In scientific and engineering computations, after performing calculations, we need to visualize numerical results for understanding and analysis. In doing so, we face the problem of choosing an efficient tool for visualization that would be easy-to-learn and convenient to use. Presently gnuplot[1] is probably the most well-known program for plotting and visualizing data. gnuplot is based on a system of commands which can be employed both interactively and in scripts. In this chapter, we consider plotting and preparing data for visualization as well as the processing of results of calculations using gnuplot. We also provide a brief description of the main features of gnuplot.

6.1 Basics

General information and the basics of working with gnuplot is discussed below. Creating scripts for gnuplot is also considered.

6.1.1 Essentials of gnuplot

The **gnuplot program** is a convenient and fairly simple tool for data visualization. It complies fully with the style and spirit of UNIX, i.e. it can do one particular thing well.

The gnuplot program is a portable open source program. It is not related to the GNU project; gnuplot uses its own license, which implies the possibility of free use and code completion in the form of plug-ins. The gnuplot project has been supported and under developed since 1986. The gnuplot program was initially created to help users visualize mathematical functions and data. Now gnuplot is used for many non-interactive tasks such as web scripting and is the visualization engine in a number of third-party applications (see e.g. Octave[2]).

The primary advantages of gnuplot are:
- actively maintained;
- stability in work;
- free and open source code;
- portability (Linux, Windows, and Mac OS);
- interactive mode or scripts;
- large set of built-in functions;
- many examples;

---

1 http://www.gnuplot.info/.
2 http://www.gnu.org/software/octave/.

- flexible and efficient programming language;
- support of many different types of output.

Before beginning a more detailed study, we should highlight that gnuplot has a flexible and convenient help system. To get help for any command, just type help <command> in an interactive session. A help topic provides complete information about the command and example of its use.

6.1.2 How to install and run gnuplot

The gnuplot program can be **installed** from the standard repository of Linux using the following command-line:

```
$ sudo apt-get install gnuplot gnuplot-x11
```

The source code can also be downloaded from the official web-site[3].

To launch gnuplot in interactive mode, we type the following command:

```
$ gnuplot
```

The program displays information about itself. Now all lines entered in the terminal are considered as commands. To quit, we can use both the quit and exit commands, or the hotkey Ctrl-D.

The welcome information of gnuplot is as follows:

```
$ gnuplot

    G N U P L O T
    Version 4.6 patchlevel 4    last modified 2013-10-02
    Build System: Linux x86_64

    Copyright (C) 1986-1993, 1998, 2004, 2007-2013
    Thomas Williams, Colin Kelley and many others

    gnuplot home:     http://www.gnuplot.info
    faq, bugs, etc:   type "help FAQ"
    immediate help:   type "help"   (plot window: hit 'h')

Terminal type set to 'unknown'
gnuplot>
```

---

3 http://www.gnuplot.info/.

6.1.3 Simple plot

We will now present a simple plotting using gnuplot. Note that the gnuplot syntax is case sensitive, i.e. plot and PLOT are not the same.

The main command of gnuplot is **plot**; it generates two-dimensional plots. It can be applied to graph functions or to visualize data from a file.

Let us begin with the simple command:

```
gnuplot> plot cos(x)
```

This command displays the graph of the function cos(x), as shown in Figure 6.1.

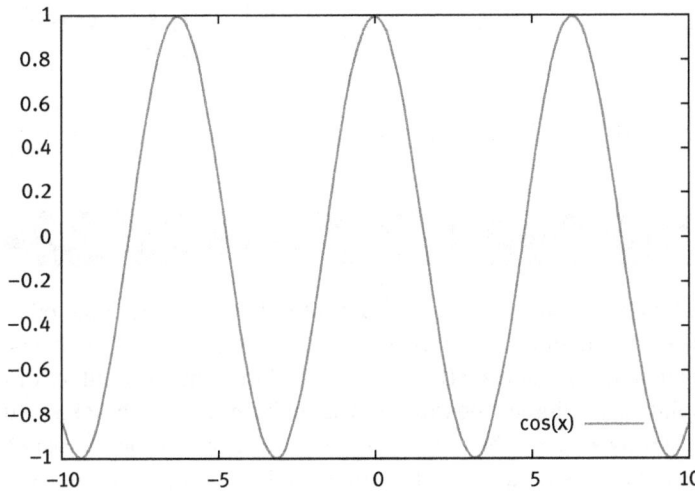

Fig. 6.1. Cosine.

Next, we add more functions and display them with the cosine. We redo the last command (using the up-arrow key or Ctrl-P) and change it to

```
gnuplot> plot cos(x), x, x**2-x
```

This command plots the cosine with the linear function x and the nonlinear function $x^2 - x$ (see Figure 6.2). The syntax of gnuplot for mathematical expressions is similar to the basic programming languages. The only exception is the sign **∗∗**; in gnuplot, this denotes the exponentiation operator.

We see that the range of y values is larger than the previous plot. Therefore, the visibility of the cosine is lost. gnuplot automatically selects a value range so that

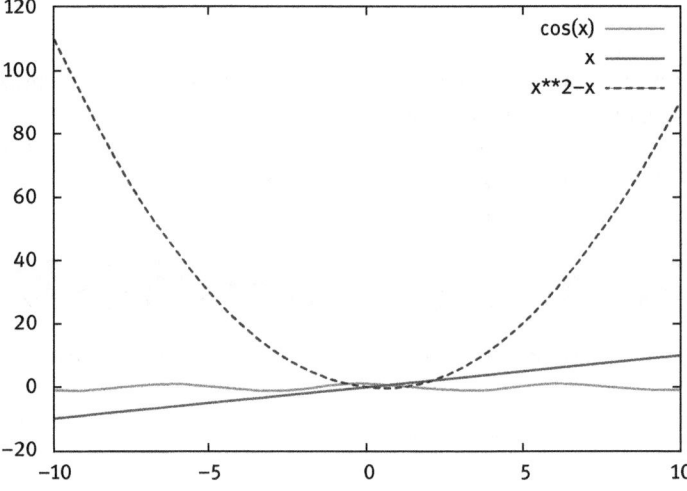

Fig. 6.2. Graphs of functions.

all values will fit. For the latter graph, we can specify new range of *y* performing the command

```
gnuplot> plot [][-2:2]  cos(x), x, x**2-x
```

This command displays the graph with the range from -2 to 2 for *y* axis, as depicted in Figure 6.3. The range is set immediately after the `plot` command in square brackets. The range of *x* is specified in the first bracket (here, we left it blank, since the values of *x* are appropriate for displaying), whereas the range of *y* is set in the second brackets. In addition, we can specify ranges using the commands `set yrange [ymin:ymax]` and `set xrange [xmin:xmax]`. Ranges must be specified before invoking the command `plot`. Thus, we can write the previous command as

```
gnuplot> set yrange [-2:2]
gnuplot> plot cos(x), x, x**2-x
```

The visualization of mathematical functions deserves a separate detailed consideration. But our main object in this study is the visualization of calculated data, and therefore, let us consider data plotting.

The `gnuplot` program reads data from a text file and the file extension may be anything. Data must be numeric and separated by tabs or spaces. If a line begins with #, then it is a comment and ignored. Let us start with a simple example. Assume that we have the file `data.txt` that contains the following:

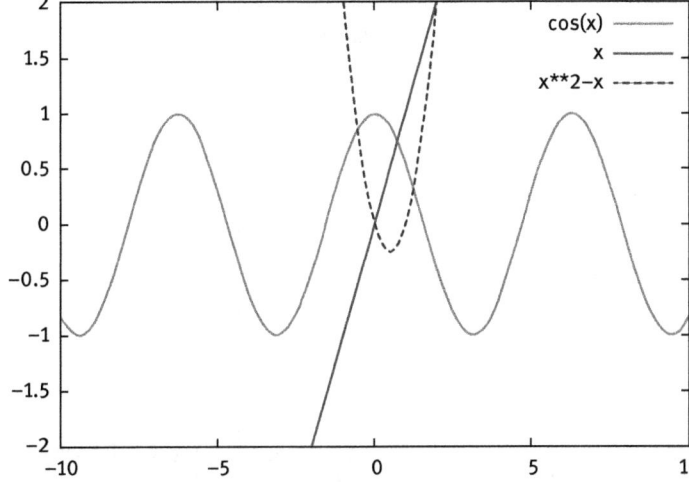

Fig. 6.3. Graphs with ranges.

#X	QX	RX
25	10	23
26	18	20
27	14	21
28	16	23
29	19	27
30	18	29
31	20	33
32	27	31
33	39	27
34	33	29
35	25	31
36	19	25
37	21	24
38	17	23
39	16	17
40	15	14
41	14	13
42	18	14

The first column is usually treated as the X value, whereas the second one is the Y value. This format is best suited for gnuplot. But gnuplot would not be so popular if it could only read this kind of data. Below we consider working with more complex file structures.

For the given file, we can visualize data using the following command:

```
gnuplot> plot "data.txt"
```

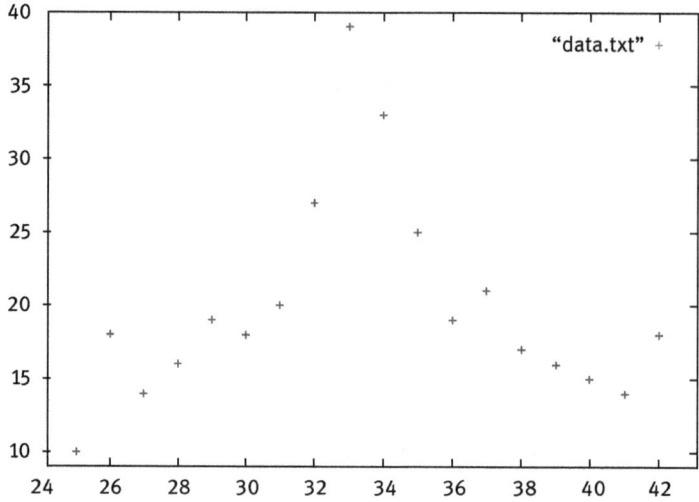

Fig. 6.4. Data visualization.

This command generates a plot (see Figure 6.4) using the first two columns, but we can specify the X and Y values using the using option of plot.

```
gnuplot> plot "data.txt" using 1:2
```

This command means that gnuplot uses the first column for X value and the second column for Y value. If the using option is not specified, then the program takes the option using 1:2 by default, and the first and second command are identical. Next, we plot the following draw:

```
gnuplot> plot "data.txt" using 1:2, "data.txt" using 1:3
```

This command generates two plots together (see Figure 6.5); for the second plot, the X and Y values are taken from the first and third column, respectively. We cannot say that this plot is clear and informative. By default, data points from the file are plotted using unconnected points with different types. We can change the plot style using the with option. This option specifies what style to use for plots. Consider the two most common styles, linespoints and lines. The style linespoints plots points connected by lines. The style lines plots lines without points. Now we change the style of the first plot to lines, the style of the second plot to linespoints. This results in the plots depicted in Figure 6.6.

```
gnuplot> plot "data.txt" using 1:2 with lines,\
"data.txt" using 1:3 with linespoints
```

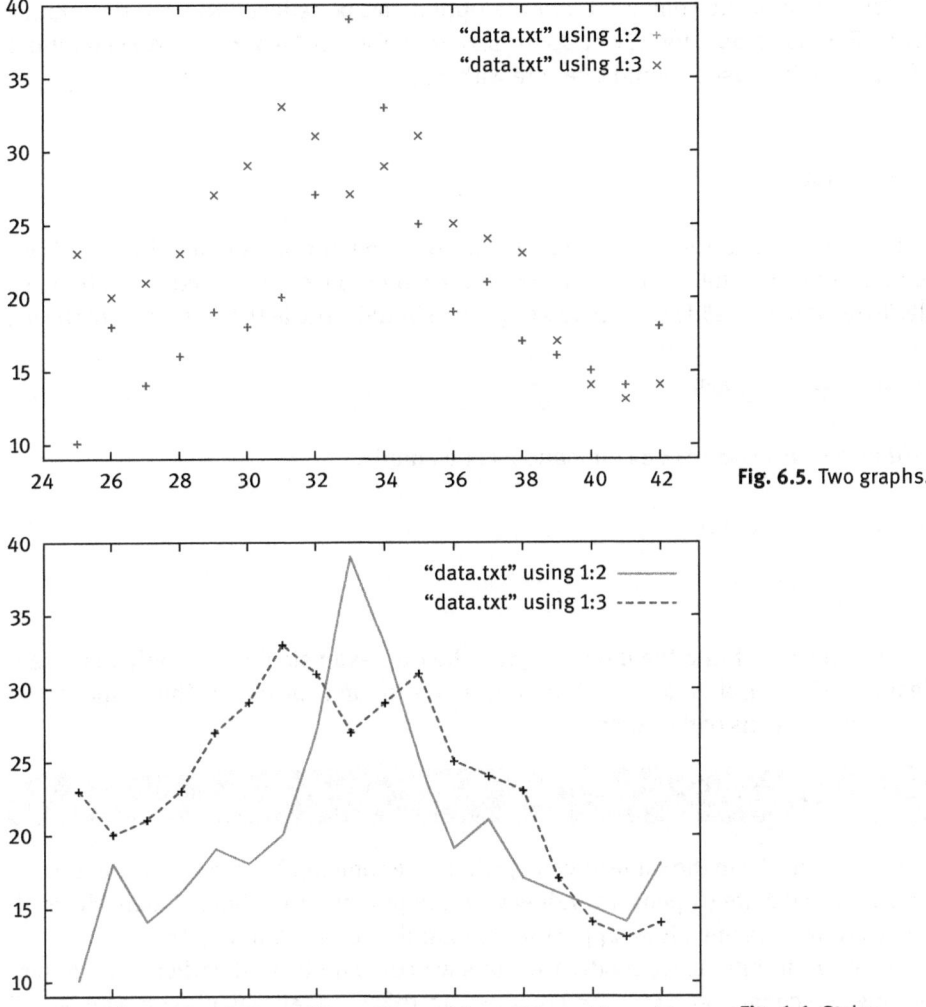

Fig. 6.5. Two graphs.

Fig. 6.6. Styles.

6.1.4 Abbreviations

The gnuplot program has a set of commands and options for each command. In addition to plot, gnuplot uses other commands (set, show, splot, load, save).

The set command can be used to specify the settings of gnuplot. The show command is employed with set and shows these settings. For example, show yrange shows the value of this option. All the settings can be shown via show all. We can reset all the settings using the reset command. To specify default values or delete an option, we need to apply unset <option>.

The name of the command and the option can be reduced to a shorter unique string. For instance, p for plot, sp for splot, t for title of plot. We cannot use s for splot because it would interfere with set.

6.1.5 Scripts

Next, we will operate with longer commands or with multiple commands. For this we need to use scripts for gnuplot. The gnuplot program executes scripts written in a file. To declare that a file contains a **script**, we should write at the beginning of the file

```
#!/usr/bin/gnuplot
```

Further, we write gnuplot commands. For example,

```
#!/usr/bin/gnuplot
set yrange [-2:2]
plot  cos(x), x, x**2-x
```

We save this script into the file ex1.gnu (the gnu extension is used only to make it clear that this script is for gnuplot; any extension may be used). The script can be executed by means of the command

```
$ gnuplot ex1.gnu
```

The script must be in the same directory where the command is invoked. If we run this script, it immediately opens a window with the plot and then immediately closes it. The window stays open if we apply the command pause -1 after plot.

Working in interactive mode, we can save commands used earlier in a file performing the command save "ex1.gnu". In this case, all default commands will also be written to the file. To load commands from the file, we use the command load "ex1.gnu". The name of the file must be inside quotes.

6.2 Data handling

The gnuplot program is well-suited for data processing. Now we will discuss commands to handle files with large data sets. Also we consider commands for changing plots.

6.2.1 Data blocks

Data sets may be more complex in comparison with the examples considered above. There are two basic cases: files with data blocks, and files containing records with multiple lines.

Consider the following case. A program performs some complicated calculations. Sometimes, it provides a summary of results, appends it to an output file and continues calculations. The resulting file is not a single data set; it is **data blocks** with many lines. Blank lines play an important role in reading data. Double blank lines indicate a separation of data blocks; a single blank line shows a break in the data.

Suppose that the resulting file has the following structure:

```
# X Y
# t=0
0        0.84
1        0.51
2        0.30
3        0.15
4        0.01


# t=1
0        0.87
1        0.35
2        0.25
3        0.16
4        0.04


# t=2
0        0.79
1        0.52
2        0.43
3        0.18
4        0.05
```

There are 3 data blocks in the above example. Each time value has its own x and y values. The index option of plot indicates a data block.

If we use the standard plot command, then it plots all 3 blocks using the same style. With index, we can get a required block, e.g.

```
gnuplot> plot "data" index 0 using 1:2 w linespoints
```

This command plots the first data block (counting from 0). The index option is employed to plot several blocks. For example, plot "data" index 2:5 plots 4 data blocks from 3 to 6.

6.2.2 Data selection

Let us consider a file that contains different records with multiple lines:

```
# experiment number - value
1          1   #procs_count
1        212   #time
2          2   #p
2        114   #t
3          4   #p
3         57   #t
4          8   #p
4         28   #t
```

In this case, gnuplot provides the every options of plot. It allows selection of certain lines or blocks of data. The general syntax of the option seems as follows:

```
every I:J:K:L:M:N

I     line increment
J     block increment
K     first line
L     first line of block
M     last line
N     last line of block
```

Here are some examples:

```
every 2        - plots every 2 lines in each block
every ::3      - plots from third line in each block
every ::3::5   - plots from 3 to 5 lines in each block
every ::0::0   - plots only the first line in each block
every 2::::6   - plots 1,3,5,7 lines in each block
every :2       - plots each 2 blocks
every :::5::8  - plots from 5 to 8 blocks
```

6.2.3 Data changing

To change data and perform simple calculations with them, there is the command using. To do this, the column number is preceded with the $ sign; this column will be changed using some mathematical operation. Such expressions must be enclosed in parentheses. The expression using 1:sqrt($2) does not work; there are no parentheses. It is possible to perform calculations with several columns. For exam-

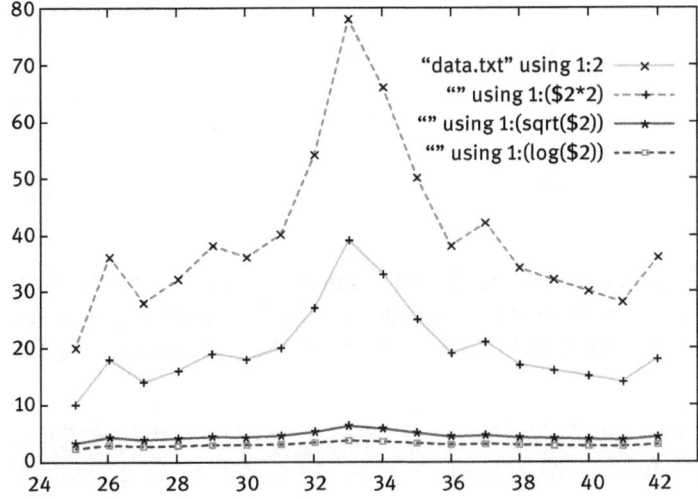

Fig. 6.7. Data changing.

ple, `using 1:2:($2*$3)` gives the product of the second and third columns. In the following example, we perform four simultaneous calculations: with y values themselves, doubled y values, the square root and the logarithm of y values (see Figure 6.7).

```
gnuplot> plot "data.txt" using 1:2 with linespoints,\
"" using 1:($2*2) with linespoints,\
"" using 1:(sqrt($2)) with linespoints,\
"" using 1:(log($2)) with linespoints
```

This method is useful when we need to translate results from one system to another, e.g. to translate temperature from Celsius to Fahrenheit.

6.3 Data interpolation

The `gnuplot` program can perform data interpolation and approximation using the `smooth` option of the `plot` command. It should be used with one of `unique`, `frequency`, `bezier`, `sbezier`, `csplines`, `acsplines` parameters.

6.3.1 Sorting and averaging data

The **smooth** option with the `unique` parameter allows us to plot unsorted data and also find the average of a set of values. If several y values exist for one x, then this option finds the average value (the `frequency` option finds the sum). We consider sorting the following data:

```
1 4
3 1
5 2
2 2 # out of order
4 3
6 3
7 1
```

Using the standard `plot` command, we get the graph presented in Figure 6.8, where the points are connected according to their order in the file. Applying `smooth unique`, we obtain the plot depicted in Figure 6.9 with the points sorted by the first column x:

```
gnuplot> plot "data3.txt" u 1:2 smooth unique with linespoints
```

Now we will show how to find the average using `smooth unique`. We have a file:

```
# X Y
# t=0
0    0.84
1    0.51
2    0.30
3    0.15
4    0.01
# t=1
0    0.87
1    0.35
2    0.25
3    0.16
4    0.04
# t=2
0    0.79
1    0.52
2    0.43
3    0.18
4    0.05
```

Here the data is not divided into blocks. We can obtain the average value of Y values for the entire period of time (see Figure 6.10) employing the following command:

```
gnuplot> plot "data4.txt" using 1:2 smooth unique with linespoints,/
"" using 1:2 with points
```

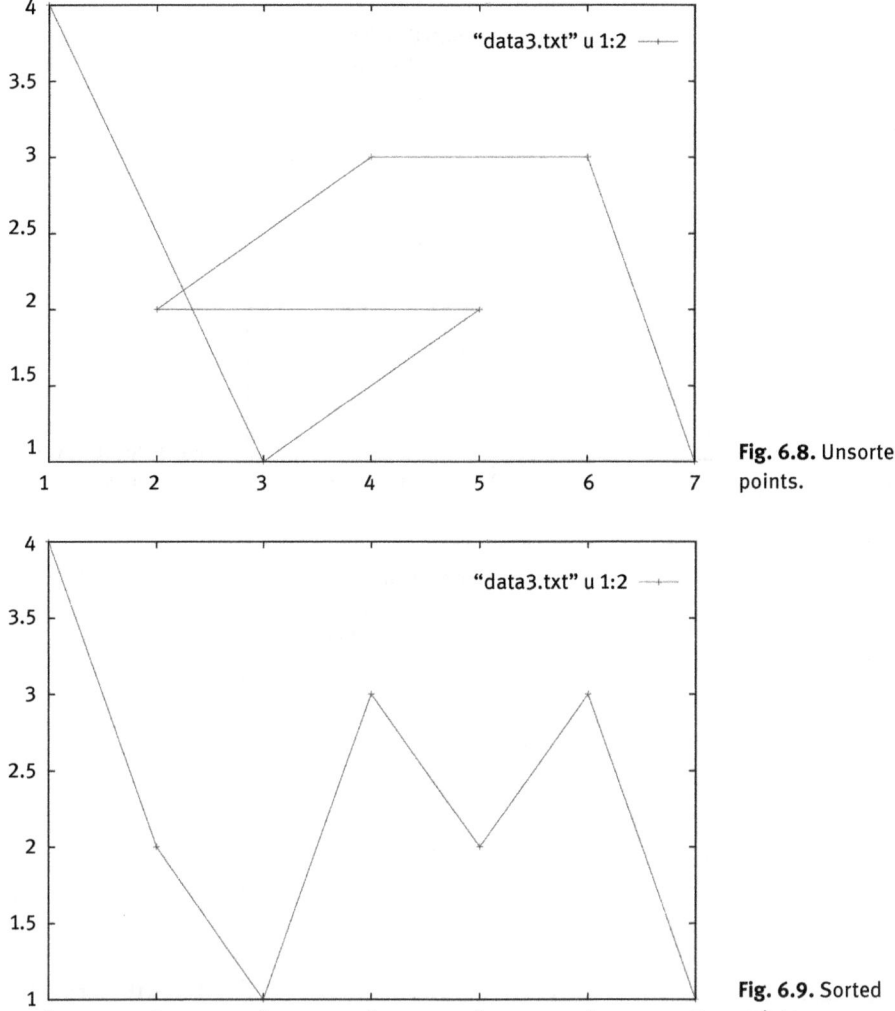

Fig. 6.8. Unsorted points.

Fig. 6.9. Sorted points.

6.3.2 Smoothing data

The bezier, sbezier, csplines, acsplines parameters allow for smoothing data:

- bezier connects points using a Bezier curve;
- sbezier first uses unique and then plots a Bezier curve;
- csplines first uses unique and then connects points using cubic splines;
- acsplines plots csplines with weights.

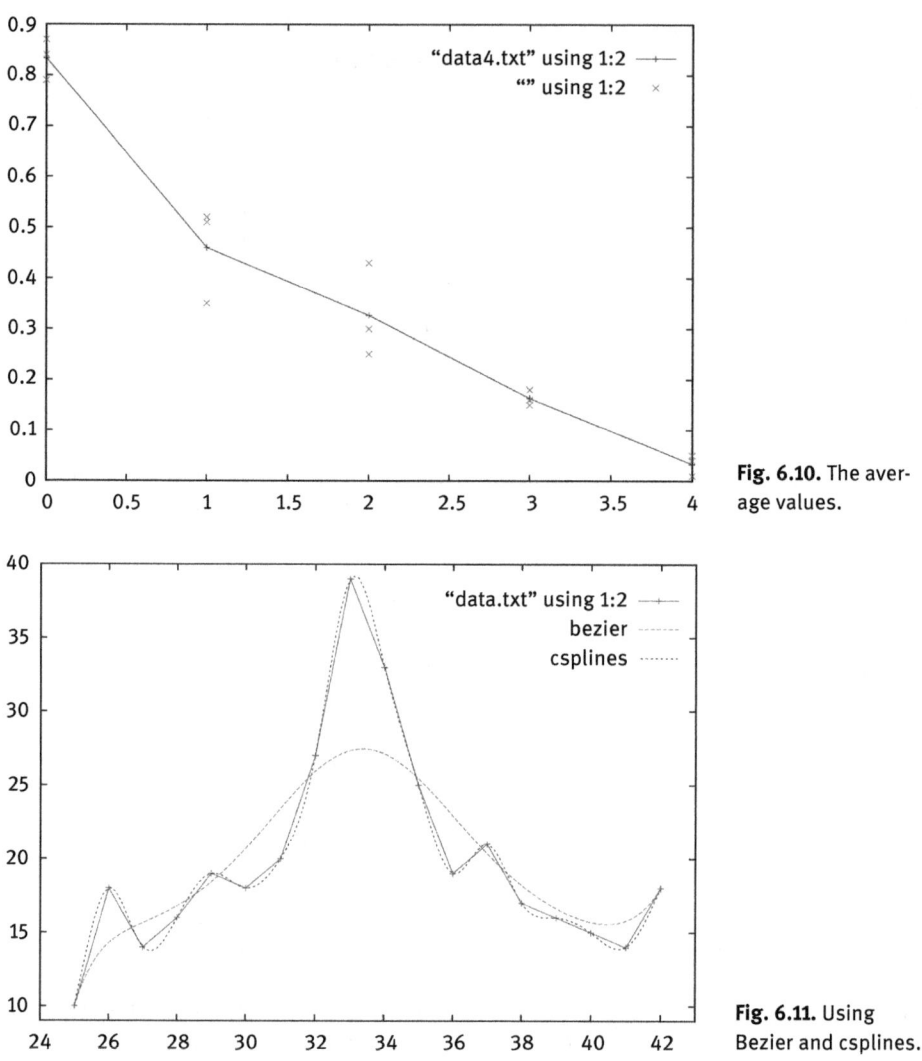

Fig. 6.10. The average values.

Fig. 6.11. Using Bezier and csplines.

Figure 6.11 shows an example of using `bezier` and `csplines`:

```
gnuplot> plot "data.txt" using 1:2 with linespoints,\
"" using 1:2 title "bezier" smooth bezier,\
"" using 1:2 title "csplines" smooth csplines
```

The following is an example of `csplines` with weights:

```
gnuplot> plot "data.txt" using 1:2 with points,\
   "" u 1:2:(1.) t "1" smooth acsplines,\
   "" using 1:2:(1/50.) t "1/50" smooth acsplines,\
   "" using 1:2:(50.) t"50" smooth acsplines,\
   "" using 1:2:(1/10000.) t "1/10000" smooth acsplines
```

In the resulting graph (see Figure 6.12), we observe that if the weights are large, then the graph tends to an interpolation curve, whereas if the weights are small, then the graph is smoother.

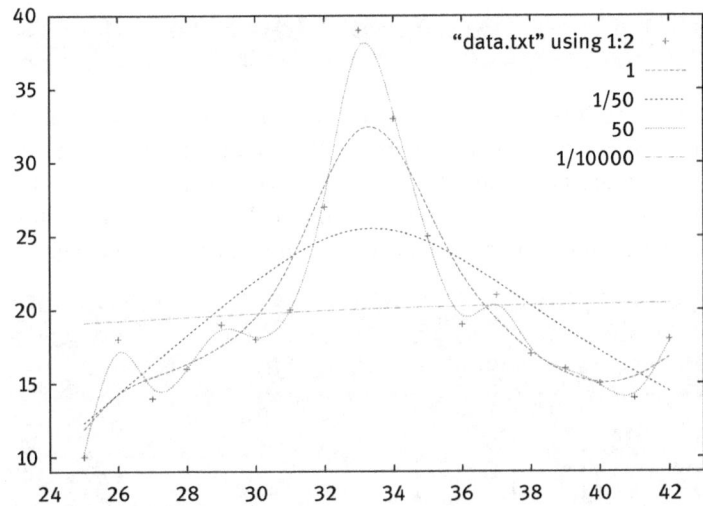

Fig. 6.12. Weights for csplines.

6.4 Using styles

We can prepare high-quality illustrations for scientific articles and presentations using gnuplot.

6.4.1 Output

To show results, gnuplot uses a terminal, which can be a device or a file. By default, gnuplot displays the results. gnuplot supports many formats for output files (ps, fig, jpeg, LaTeX, metafont, pbm, pdf, png, postscript, svg). It is possible to change the **type of the terminal** using the following command:

```
gnuplot> set terminal <type>
```

A list of possible types:
- `windows` displays in the terminal of `Windows OS`;
- `X11` displays in the terminal of `Linux OS`;
- `png` displays in a `png` file (raster format);
- `jpeg` displays in a `jpeg` file (raster format);
- `postscript` displays in the vector format;
- `latex` displays in the LaTeX format.

We can see the full list of available terminals via the command

```
gnuplot> set terminal
```

Let us study the `postscript` format, which is a vector format and is better suited for printing and presentation in scientific publications. By default, this terminal sets to monochrome coloring, which is suitable for black and white printing. For color graphs, we can employ the `postscript color` option or the `png` and `jpeg` formats. If we specify a terminal, the default options of the terminal are printed, e.g.

```
gnuplot> set terminal postscript
Terminal type set to 'postscript'
Options are 'landscape noenhanced defaultplex \
   leveldefault monochrome colortext \
   dashed dashlength 1.0 linewidth 1.0 butt noclip \
   palfuncparam 2000,0.003 \
   "Helvetica" 14 '
```

Here the font is Helvetica at 14pt, the orientation is `landscape`, color is monochrome, the text is without superscripts and subscripts (`noenhanced`), etc.

The terminal capabilities can be seen using the `test` command. For example, Figure 6.13 is the test window of the `postscript` terminal with the `color` parameter:

```
gnuplot> set terminal postcript color
gnuplot> set output "test.ps"
gnuplot> test
```

Here we observe the new command `set output`. This command specifies the file to which we want to output the results.

The available styles for lines and points are shown at the right of Figure 6.13. Styles may be different for each terminal.

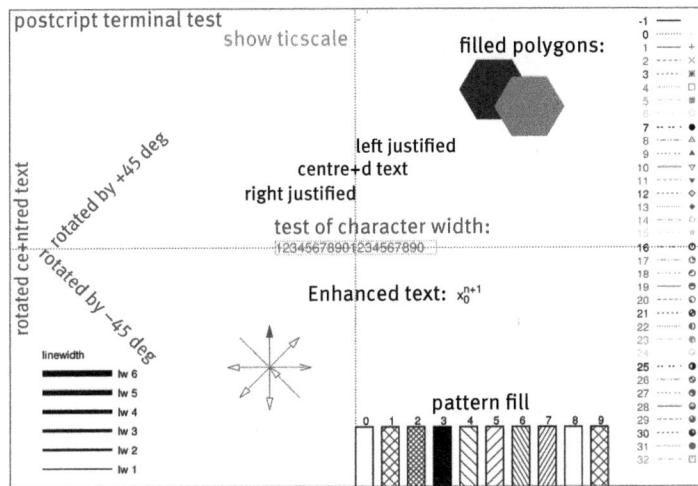

Fig. 6.13. Capabilities of postscript.

6.4.2 Styles

A graph can be displayed using one of the following **styles**: lines, points, linespoints, impulses, dots, steps (fsteps, histeps), errorbars (or yerrorbars), xerrorbars, xyerrorbars, boxes, boxerrorbars, or boxxyerrorbars, histograms.

The styles can be set directly for a function plot or globally using the following commands:

```
set style function <style>
set style data <style>
```

The lines style displays points connected by lines. The line width and type are specified by the options linewidth and linetype, respectively.

The points style displays unconnected symbols. The symbol type is specified by the pointtype option (or pt); its size is given by the pointsize option (or ps).

The linespoints style is the combination of lines and points and displays symbols connected by lines. The parameters of this style are linewidth, linetype, pointtype, and pointsize.

The impulses style displays vertical lines from x axis to points, as shown in Figure 6.14.

The boxes style is necessary for two-dimensional plots. It draws a box from the x axis to the given point (see Figure 6.15).

The dots style plots a small dot at each point.

The steps style connects consecutive points with horizontal and vertical line segments, i.e. we obtain ladder (see Figure 6.16).

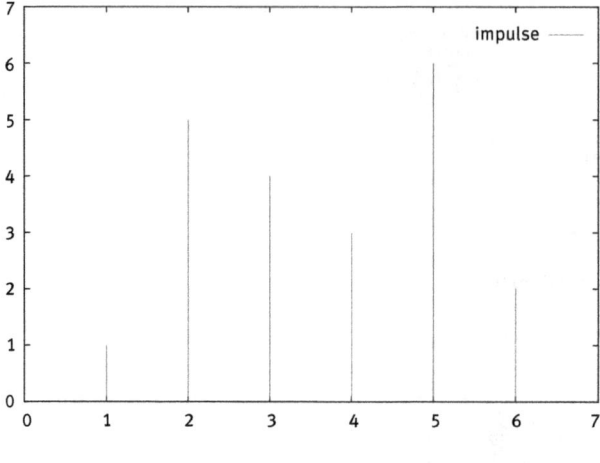

Fig. 6.14. Impulses.

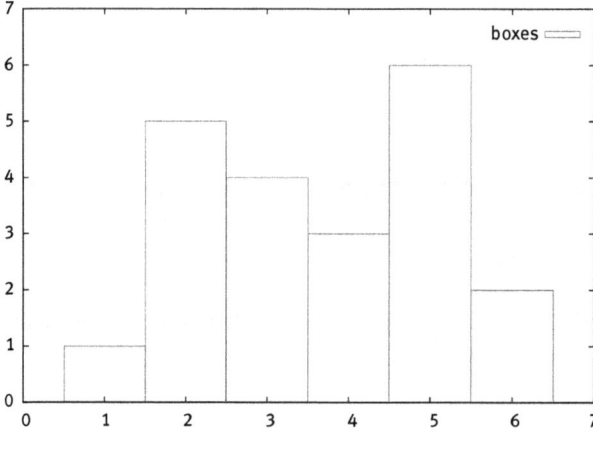

Fig. 6.15. Boxes.

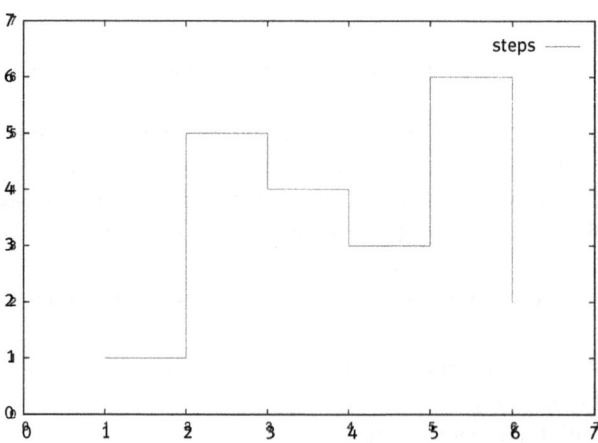

Fig. 6.16. Steps.

The fsteps style is similar to the previous one. The difference is in the order of a ladder drawing (Figure 6.17 presents this case).

The histeps style draws ladder so that the middle of step is a point. We can observe this situation in Figure 6.18.

The histograms style obviously draws a histogram. In this case, we can specify additional options. Histograms can be built in different ways: box or impulses. But histograms is most convenient.

The standard histogram (histogram clustered) view shows several columns next to each other (see Figure 6.19).

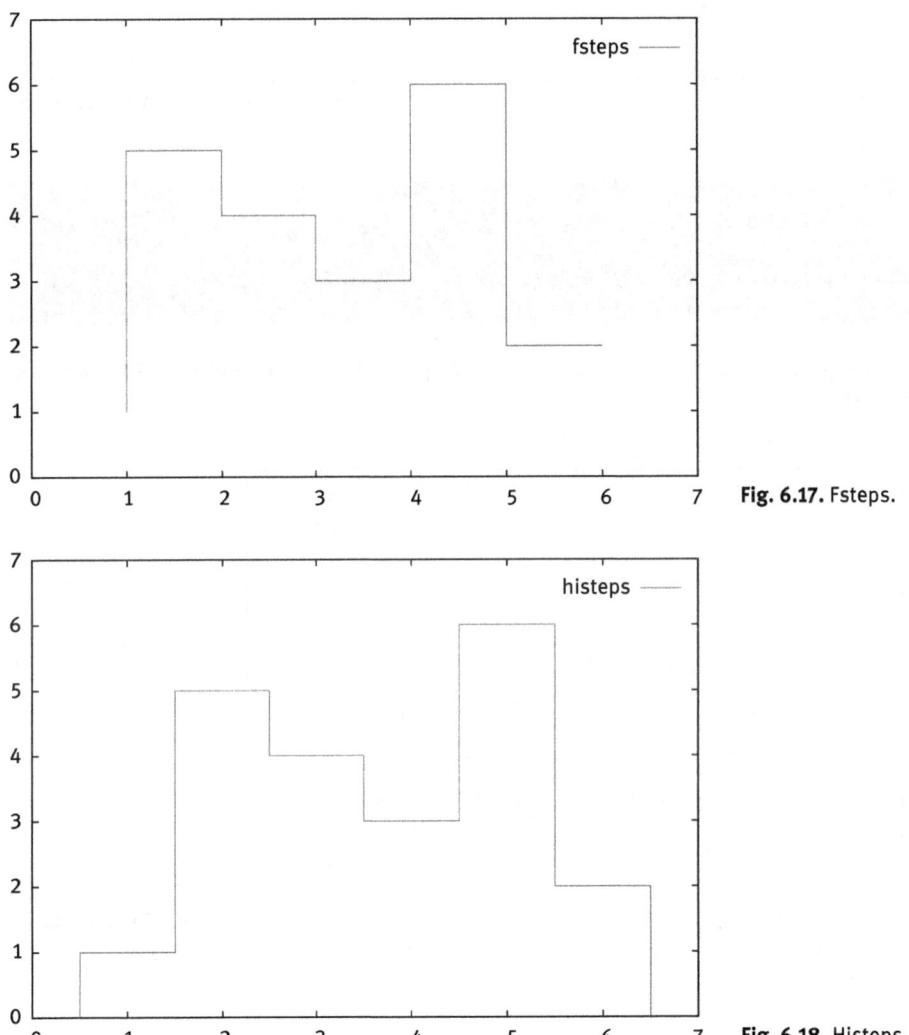

Fig. 6.17. Fsteps.

Fig. 6.18. Histeps.

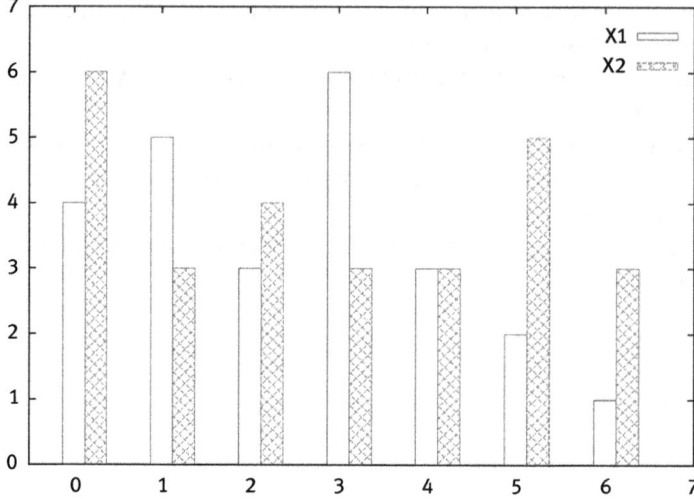

Fig. 6.19. Histogram.

```
gnuplot> set style fill pattern
gnuplot> set style histogram clustered
gnuplot> set style data histograms
gnuplot> plot "data5.txt" u 2 t "X1", "" u 3 t "X2"
```

For example, `histogram rowstacked` displays values in one column (see Figure 6.20).

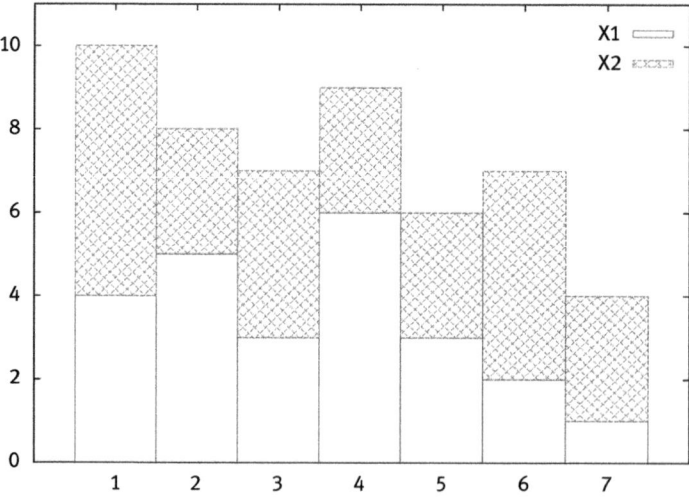

Fig. 6.20. Stacked histogram.

```
gnuplot> set style fill pattern
gnuplot> set style histogram rowstacked
gnuplot> set style data histograms
gnuplot> plot "data5.txt" u 2:xtic(1) t "X1", "" u 3 t "X2"
```

6.4.3 Line style

To set the **style of lines**, we use inline styles using the option `lt <style_number>` (`linetype`), e.g.:

```
gnuplot> plot [][-2:2] cos(x) lt 6, x linetype 10
```

Figure 6.21 presents numbering, lines, and symbols styles.

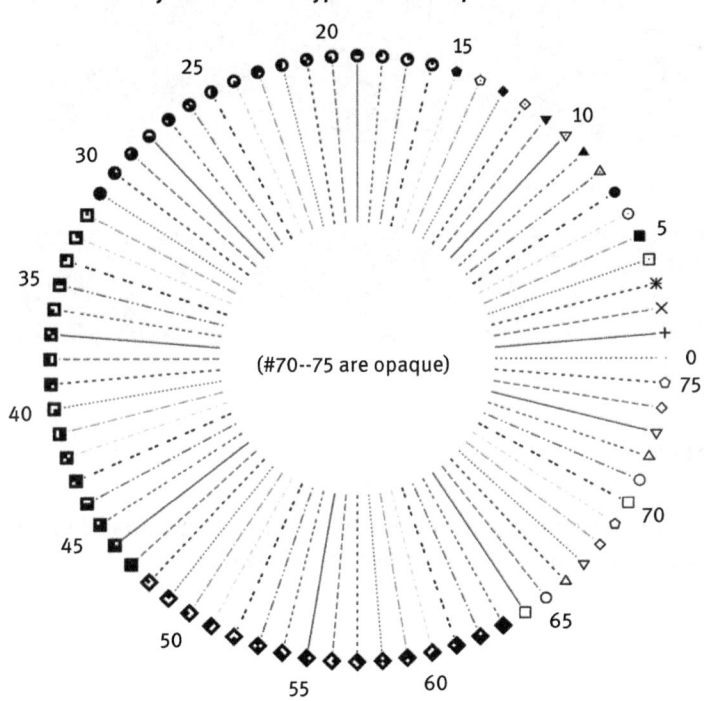

Fig. 6.21. Lines and symbols in postscript.

To specify options of lines, we apply two-letter abbreviations which begin with l. For example, lw is linewidth. A list of abbreviations can be seen using the following command:

```
gnuplot> help set style line
```

This command prints the options supported by the command set style line. Note that the options separated be a dash | are equivalent, i.e. the linecolor option can be written as the lc abbreviation.

```
. . .
 Syntax:
   set style line <index> default
   set style line <index> {{linetype  | lt} <line_type> | <colorspec>}
                          {{linecolor  | lc} <colorspec>}
                          {{linewidth  | lw} <line_width>}
                          {{pointtype  | pt} <point_type>}
                          {{pointsize  | ps} <point_size>}
                          {{pointinterval | pi} <interval>}
                          {palette}
   unset style line
   show style line
. . .
```

For manual tuning, we use the following options:

```
gnuplot> set style line 1 lt 3 lw 2
gnuplot> set style line 2 lt 3 lw 4
gnuplot> plot "data" u 1:2 w l ls 1, "data" u 1:3 w l ls 2 lc "green"
```

6.5 Graph decoration

For greater clarity, it is sometimes necessary to somehow **decorate graphs**. Here we consider how to modify and customize legends, titles, labels, and axes.

6.5.1 Legend

We have already discussed how to change the **legend** using title "name". Here we discuss how to remove the legend and change its location.

There are two ways to remove the legend. The first way is implemented by the command

```
gnuplot> unset key
```

The second method is to apply the `notitle` option of the `plot` command. In the following command, file data has a legend, but the legend is cancelled:

```
gnuplot> plot f(x) notitle, "file.dat" title "data"
```

The legend is usually displayed at the upper right of a graph. Its location can be changed using the command `set key`. If we employ the command

```
gnuplot> set key left bottom
```

then the legend appears at the lower left of the graph. We can also set the location of the legend directly. If we want to place it at the position $(X, Y) = (100, 100)$, then we write

```
gnuplot> set key 100,100
```

The coordinate system is defined by the X and Y axes. To customize the coordinate system, it is necessary to work with the help `help coordinates`.

6.5.2 Axes

Sometimes we need to place several graphs in one figure, where *x* and *y* values may differ. Using `gnuplot`, it is easy to specify values of each **axis** independently. By default, the Y2 axis is the same as the Y axis. For example, we make them different and plot `cos(x)` and the square of this function in the same graph. The axis can be selected using `x1y1, x1y2, x2y1, x2y2`. Here axis `x1y2` means that the function will be scaled to fit the primary X axis and the secondary (right) Y axis. To customize the secondary Y axis, we apply the `nomirror` option for `set ytics`. We specify `set y2tics` to control tics on the Y2 axis:

```
gnuplot> set xrange [0:2*pi]
gnuplot> set yrange [-1:1]
gnuplot> set y2range [0:1]
gnuplot> set y2tics 0, 0.2
gnuplot> set ytics nomirror
gnuplot> plot cos(x) axis x1y1, cos(x)**2 axis x1y2
```

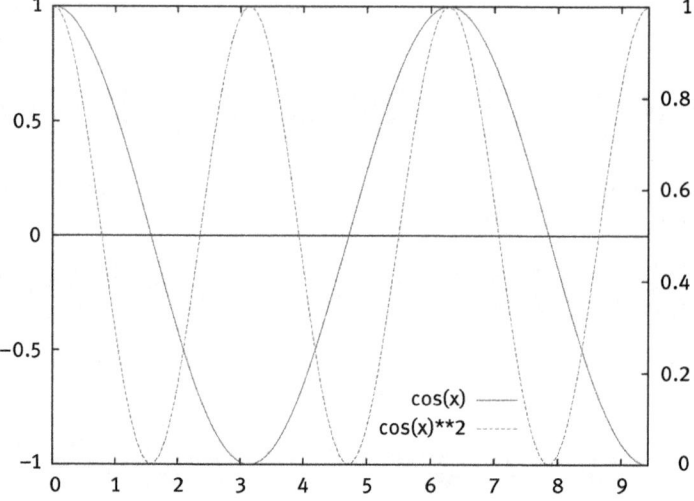

Fig. 6.22. The secondary Y axis.

This command draws the graph shown in Figure 6.22.

To control tic marks on axes, we use the commands `set xtics`, `set ytics`, `set mxtics`, `set mytics`. For example,

```
gnuplot> set xtics # tics every 5 points
gnuplot> unset ytics # cancel tics on the Y axis
gnuplot> set ytics 3, 5, 23 # tics from  3 to 23 every 5;
gnuplot> set mxtics 10 # divide every  xtics-interval into
         10 sub-intervals;
```

By default, `gnuplot` draws all four axes, but sometimes we need to make some axes invisible. To remove undesired axes, the `set border n` command is employed. Each axis is represented by an integer number. The bottom axis is 1, the left axis is 2, the top axis is 4 and the right axis is 8, n is the sum of numbers assigned to the axes which will be built.

To disable tic marks, we use the commands `set no{x|y}tics` or `set {x|y}tics nomirror`.

For example, if we want to draw a graph without the top and right axes, then we perform the following commands:

```
gnuplot> set border 3
gnuplot> set xtics nomirror
gnuplot> set ytics nomirror
```

An example of such a graph is depicted in Figure 6.23.

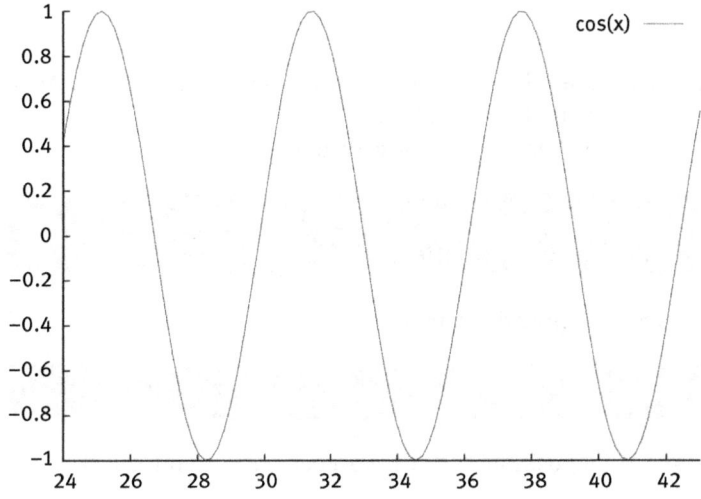

Fig. 6.23. Graph without the right and top axes.

If we want to have an axis passing through the zero coordinates, then we use the command `set {x|y}zeroaxis`. By default, the zero axis is drawn by points, but it is easy to redefine this via the commands `line_style`, `line_type`, `line_width`. For instance, we can do this as follows:

```
gnuplot> set xzeroaxis lt -1
gnuplot> set yzeroaxis
```

This command draws the graph presented in Figure 6.24.

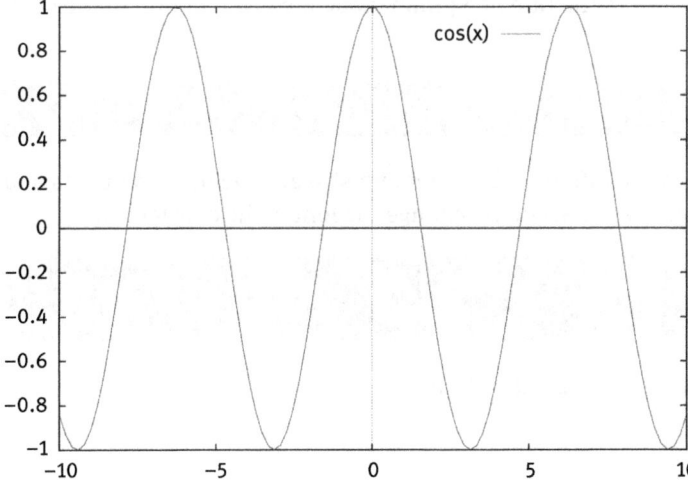

Fig. 6.24. Zero axis.

6.5.3 Labels

It is possible to change the axis **labels** and append additional text in a graph. Also, we can rotate labels as well as set their font and size.

The following is an example of employing these commands:

```
gnuplot> set ylabel "label_y"
gnuplot> set xlabel "label_x" font "Helvetica,18"
```

To put the label at any place, we apply the command

```
gnuplot> set label "text" at 0,0
```

This command puts `"text"` 0 at position 0,0 with respect to the X and Y axes.

6.5.4 Font

By default, gnuplot uses the Helvetica font at 14pt. But sometimes, we need graphs with different fonts. To change the font in the axis labels and legend, we use the command

```
gnuplot> set terminal postscript enhanced "Courier" 20
```

The quotation marks around the name of the font are required. The size is specified in pixels after the font.

We can change the font in the title of a graph using the font option of the command set title:

```
gnuplot> set title "Some title" font "Times-Roman,40"
```

Note that this syntax is quite different. The font and size are enclosed in quotes and separated by the comma. We can similarly change the font of the axis labels:

```
gnuplot> set xlabel "Xlabel" font "Courier,20"
gnuplot> set ylabel "Ylabel" font "Courier,20"
```

Now a graph will be drawn with specified fonts.

6.5.5 Grid

We can draw **grid lines** for minor tic marks specifying

```
gnuplot> set grid xtics ytics mxtics mytics
```

Here we should follow how gnuplot defines minor tic marks. If they are determined wrong, then they should be directly specified. For example,

```
gnuplot> set mxtics 5
gnuplot> set mytics 5
```

On the other hand, if we need to remove grid lines for minor or major tic marks, we employ the following command:

```
gnuplot> set grid noxtics noytics mxtics mytics
```

Aleksandr G. Churbanov, Alexandr E. Kolesov, and
Petr N. Vabishchevich

7 Mathematical modeling

Abstract: Examples of the numerical solution of model problems from various fields of science and engineering are presented in this chapter. All research begins with the mathematical formulation of a problem and the selection of a numerical method. The programs developed here are implemented using the GSL scientific library. The numerical results are visualized employing gnuplot.

7.1 Approximation of experimental data

The problem of constructing an approximate function for a set of one-dimensional data is formulated and solved. The polynomial approximation including the least squares fitting of polynomials is considered.

7.1.1 Problem formulation

Assume that values of a function $y_i = f(x_i)$, $i = 0, 1, \ldots, n$ are known at points $x_0 < x_1 < \cdots < x_n$ distributed within an interval $[a, b]$, e.g. we have the measurement data. Let us consider the problem of constructing the function $f(x)$ which approximates this data.

Define on the interval $[a, b]$ a system of functions $\{\phi_k(x)\}_{k=0}^{p}$, and introduce the approximating function

$$\phi(x) = \sum_{k=0}^{p} c_k \phi_k(x) \tag{7.1}$$

with real coefficients c_k, $k = 0, 1, \ldots, p$ ($p \leq n$). In the case of **polynomial approximation**, we have

$$\phi_k(x) = x^k, \quad k = 0, 1, \ldots, p.$$

7.1.2 Least squares method

In equation (7.1), the coefficients c_k, $k = 0, 1, \ldots, p$ can be determined in various ways. Using the **least squares method**, these coefficients are evaluated by minimizing the sum of the squared differences between the approximating function and the experimental values:

$$c_k, k = 0, 1, \ldots, p : \quad \min \sum_{i=0}^{p} (\phi(x_i) - y_i)^2.$$

Define a p-dimensional vector of unknown coefficients $\boldsymbol{c} = \{c_k\} = \{c_1, c_2, \ldots, c_p\}$ and let $\boldsymbol{f} = \{\phi(x_i)\} = \{\phi(x_1), \phi(x_2), \ldots, \phi(x_n)\}$. According to equation (7.1), we get

$$\boldsymbol{f} = R\boldsymbol{c},$$

where R is the rectangular matrix $R = \{r_{ik}\}$ with the elements $r_{ik} = \phi_k(x_i), i = 0, 1, \ldots, n$, $k = 0, 1, \ldots, p$.

7.1.3 Program

For a polynomial approximation based on the least squares method, we apply the function `gsl_multifit_linear` from GSL. The code implementing this algorithm is as follows:

Listing 7.1.

```
1   // file: bvpt.cpp
2   #include <iostream>
3   #include <iomanip>
4   #include <fstream>
5   #include <math.h>
6   #include <gsl/gsl_multifit.h>
7   using namespace std;
8   int main() {
9       const int n = 20;
10      const int p = 3;
11      gsl_vector * x = gsl_vector_alloc(n + 1); // x
12      gsl_vector * y = gsl_vector_alloc(n + 1); // y
13      gsl_vector * c = gsl_vector_alloc(p + 1); // c
14      gsl_multifit_linear_workspace * ws =
15          gsl_multifit_linear_alloc(n + 1, p + 1);
16      gsl_matrix * R = gsl_matrix_alloc(n + 1, p + 1);
17      gsl_matrix * cov = gsl_matrix_alloc(p + 1, p + 1);
18      double chisq;
19      // data
20      int i, k;
21      double a = 0.0;
22      double b = 3.0;
23      for (i = 0; i < n + 1; i++) {
24          double xi = i * (b - a) / n;
25          double yi = xi * sin(xi);
26          gsl_vector_set(x, i, xi);
27          gsl_vector_set(y, i, yi);
28      }
29      // R matrix
30      for (i = 0; i < n + 1; i++) {
31          gsl_matrix_set(R, i, 0, 1.0);
```

```
32        for (k = 0; k < p + 1; k++) {
33            gsl_matrix_set(R, i, k, pow(gsl_vector_get(x, i), k));
34        }
35    }
36    // solution
37    ws = gsl_multifit_linear_alloc(n + 1, p + 1);
38    gsl_multifit_linear(R, y, c, cov, &chisq, ws);
39    // save and print
40    ofstream file("mnk.dat");
41    file << setw(12) << "#x" << " " << setw(12) << "y" << " "
42        << setw(12) << "f" << endl;
43    for (i = 0; i < n + 1; i++) {
44        file << scientific << setprecision(5) << setw(12)
45            << gsl_vector_get(x, i) << " " << setw(12)
46            << gsl_vector_get(y, i) << endl;
47    }
48    for (i = 0; i < n + 1; i++) {
49        double fi = 0.0;
50        for (k = 0; k < p + 1; k++) {
51            fi += gsl_vector_get(c, k) * pow(gsl_vector_get(x, i), k);
52        }
53        file << scientific << setprecision(5) << setw(12)
54            << gsl_vector_get(x, i) << " " << setw(12)
55            << gsl_vector_get(y, i) << " " << setw(12)
56            << fi << endl;
57    }
58    for (k = 0; k < p + 1; k++) {
59        printf("c(%u) = %g\n", k, gsl_vector_get(c, k));
60    }
61    gsl_multifit_linear_free(ws);
62    gsl_matrix_free(R);
63    gsl_matrix_free(cov);
64    gsl_vector_free(x);
65    gsl_vector_free(y);
66    gsl_vector_free(c);
67    return 0;
68  }
```

In this example, the values of the function $x \sin(x)$ at 21 points on the interval $[0, 4]$ are treated as the observation data for approximation.

7.1.4 Computations

The results of computations (the values of the approximating function at observation points) are written in the file mnk.dat. The gnuplot script (file mnk.gnu) for visualization is also given.

```
#!/usr/bin/gnuplot
set terminal png
set output "mnk.png"
set xlabel "x"
plot "mnk.dat" using 1:2 title "y" with points, \
"mnk.dat" using 1:3 title "f" with lines
pause -1
```

This program produces the following coefficients of the approximating polynomial:

```
c(0) = -0.0174225
c(1) = 0.0146268
c(2) = 1.26702
c(3) = -0.409814
```

The input data and the resulting function are plotted in Figure 7.1.

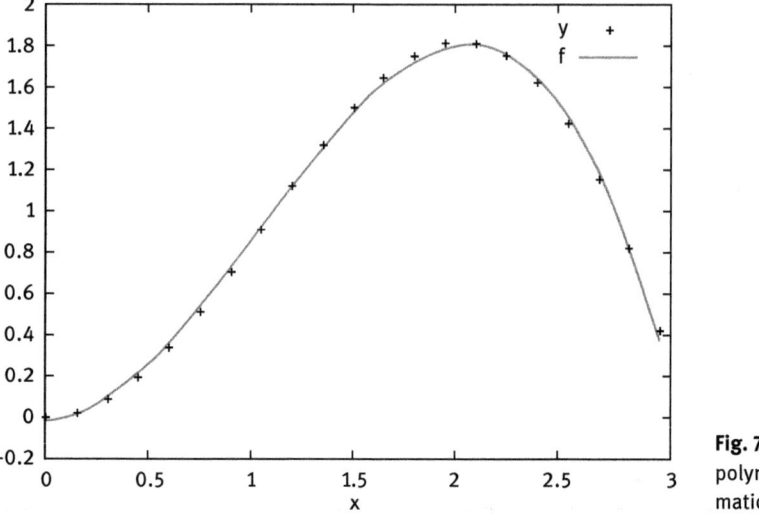

Fig. 7.1. Third-degree polynomial approximation.

The second-degree polynomial approximation yields the coefficients

```
c(0) = -0.490451
c(1) = 2.17414
c(2) = -0.577144
```

The corresponding plot is depicted in Figure 7.2.

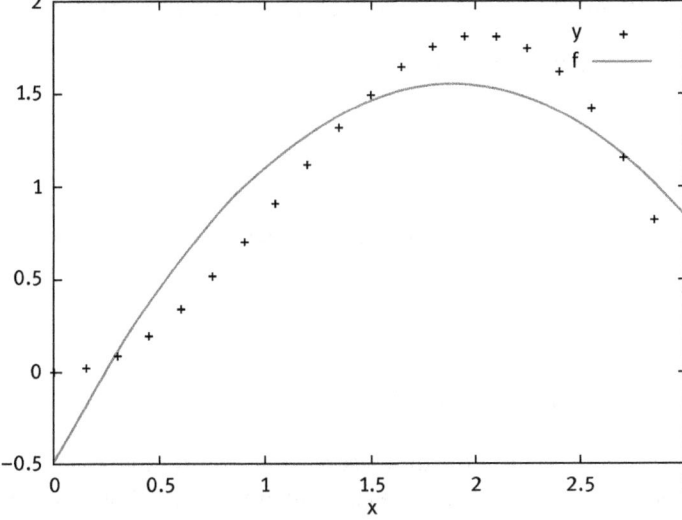

Fig. 7.2. Second-degree polynomial approximation.

7.2 Predator-prey system

A numerical study of the predator-prey system based on the numerical solution of the **Lotka–Volterra equations** is presented below.

7.2.1 Mathematical model

We consider two species of animals that inhabit a certain area. Rabbits (prey) feed on plants, and foxes (predators) hunt for rabbits. Let v be the number of rabbits and w the number of foxes. The behavior of the system is governed by the Lotka–Volterra equations:

$$\frac{dv}{dt} = (\alpha_1 - \alpha_2 w)v, \tag{7.2}$$

$$\frac{dw}{dt} = (-\alpha_3 + \alpha_4 v)w, \tag{7.3}$$

where t is time.

The number of prey per unit of time is increased due to the birth of new prey (the reproduction rate by the number of prey), and it is decreased due to prey being eaten. Accordingly, in equation (7.2), the coefficient α_1 is the growth rate of the prey population (birthrate), and α_2 stands for the coefficient of predation on a prey (the probability that a prey meets a predator and will be eaten). The growth of the predator population per unit of time is proportional to the quality of the food, and the loss is due to natural death. In equation (7.3), α_3 is the coefficient of predator mortality, and α_4 ensures the growth of the population by eating prey.

The dynamics of the system (7.2), (7.3) is determined by the initial conditions

$$v(0) = v^0, \quad w(0) = w^0. \tag{7.4}$$

7.2.2 Stationary state

We can define the stationary state of the system (7.2), (7.3). The conditions

$$\frac{dv}{dt} = 0, \quad \frac{dw}{dt} = 0$$

are fulfilled if

$$\bar{w} = \frac{\alpha_1}{\alpha_2}, \quad \bar{v} = \frac{\alpha_3}{\alpha_4}.$$

It seems reasonable that in some range of the initial conditions the number of species will become stable. However, a significantly more complex unsteady behavior of the above-mentioned nonlinear dynamic system is also observed.

7.2.3 Numerical method

For the numerical study of the mathematical model of predator-prey, the Runge–Kutta method with an adaptive step (**Runge–Kutta–Fehlberg method**) from GSL is applied.

In numerically solving the Cauchy problem for the differential equations, the Runge–Kutta method with the evaluation of an integration step is used. The simplest strategy of changing an integration step is based on solving the problem with two time steps. The second one is more interesting and involves the solution of the problem with a single step, but using methods with different accuracy.

The Cauchy problem for the system of ODEs under the consideration appears to be

$$\frac{du_i(t)}{dt} = f_i(t, u_1, u_2, \ldots, u_m), \quad t > 0,$$

$$u_i(0) = u_i^0, \quad i = 1, 2, \ldots, m.$$

In vector notation, this problem may be rewritten as the Cauchy problem for the single equation

$$\frac{du(t)}{dt} = f(t, u), \quad t > 0,$$

$$u(0) = u^0,$$

where $u = \{u_1, u_2, \ldots, u_m\}$ is the vector of unknowns, and $f = \{f_1, f_2, \ldots, f_m\}$ is the right-hand side vector.

In **one-step Runge–Kutta methods,** the transition from the time level t^n to the next level $t^{n+1} = t^n + \tau$ may be written as

$$\frac{y^{n+1} - y^n}{\tau} = \sum_{i=1}^{s} b_i k_i,$$

where

$$k_i = f(t^n + c_i \tau, y^n + \tau \sum_{j=1}^{s} a_{ij} k_j), \quad i = 1, 2, \ldots, s.$$

The computational implementation is based on s computations of the function f (the s-stage algorithm).

In the explicit Runge–Kutta–Fehlberg method, we have

$$k_1 = f(t^n, y^n),$$

$$k_2 = f\left(t^n + \frac{1}{4}\tau, y^n + \frac{1}{4}\tau k_1\right),$$

$$k_3 = f\left(t^n + \frac{3}{8}\tau, y^n + \frac{3}{32}\tau k_1 + \frac{9}{32}\tau k_2\right),$$

$$k_4 = f\left(t_n + \frac{12}{13}\tau, y^n + \frac{1932}{2197}\tau k_1 - \frac{7200}{2197}\tau k_2 + \frac{7296}{2197}\tau k_3\right),$$

$$k_5 = f\left(t_n + \tau, y^n + \frac{439}{216}\tau k_1 - 8\tau k_2 + \frac{3680}{513}\tau k_3 - \frac{845}{4104}\tau k_4\right),$$

$$k_6 = f\left(t_n + \frac{1}{2}\tau, y^n - \frac{8}{27}\tau k_1 + 2\tau k_2 - \frac{3544}{2565}\tau k_3 + \frac{1859}{4104}\tau k_4 - \frac{11}{40}\tau k_5\right).$$

The approximate solution is determined with fourth-order accuracy according to

$$\frac{y^{n+1} - y^n}{\tau} = \frac{25}{216}k_1 + \frac{1408}{2565}k_3 + \frac{2197}{4101}k_4 - \frac{1}{5}k_5.$$

To control accuracy, we employ the numerical solution $\bar{y}^{n+1}$, which is evaluated from

$$\frac{\bar{y}^{n+1} - y^n}{\tau} = \frac{16}{135}k_1 + \frac{6656}{12825}k_3 + \frac{28561}{56430}k_4 - \frac{9}{50}k_5 + \frac{2}{55}k_6$$

and corresponds to the method of fifth-order accuracy.

7.2.4 Program

To solve numerically the Cauchy problem for the system (7.2)–(7.4), the Runge–Kutta–Fehlberg method of fourth-fifth order accuracy mentioned above is employed. This method is implemented by means of GSL and is presented in the following code:

Listing 7.2.

```cpp
1   //   file: lotka.cpp
2   #include <iostream>
3   #include <iomanip>
4   #include <fstream>
5   #include <gsl/gsl_errno.h>
6   #include <gsl/gsl_matrix.h>
7   #include <gsl/gsl_odeiv.h>
8   using namespace std;
9   // function prototypes
10  int rhs(double t, const double y[], double f[], void *params);
11  // main program
12  int main() {
13      int dim = 2; // number of differential equations
14      double al[4] = {0.05, 0.005, 0.01, 0.0002}; // parameters
15      double y[2]; // current solution vector
16      double t, tNext; // current and next independent variable
17      double tau, tau0 = 1e-6; // step size, starting step size
18      double tMin, tMax; // range of t and step size for output
19      double epsA = 1.e-8; // absolute error requested
20      double epsR = 1.e-10; // relative error requested
21      // Runge-Kutta-Fehlberg (4,5) method
22      const gsl_odeiv_step_type *type = gsl_odeiv_step_rkf45;
23      gsl_odeiv_step *step = gsl_odeiv_step_alloc(type, dim);
24      gsl_odeiv_control *control = gsl_odeiv_control_y_new(epsA, epsR);
25      gsl_odeiv_evolve *evolve = gsl_odeiv_evolve_alloc(dim);
26      gsl_odeiv_system system; // system structure
27      system.function = rhs; // the right-hand-side functions
28      system.dimension = dim; // number of diffeq's
29      system.params = &al; // parameters to pass to rhs
30      tMin = 0.; // starting t value
31      tMax = 500.; // final t value
32      tau = 1.;
33      // initial data
34      y[0] = 10.;
35      y[1] = 10.;
36      t = tMin; // initialize t
37      // save initial values
38      ofstream file("lotka.dat");
39      file << setw(12) << "#t" << " " << setw(12) << "y[0]" << " "
40          << setw(12) << "y[1]" << endl;
41      file << scientific << setprecision(5) << setw(12) << t << " "
42          << setw(12) << y[0] << " " << setw(12) << y[1] << endl;
43      // step to tmax from tmin
44      for (tNext = tMin + tau; tNext <= tMax; tNext += tau) {
45          while (t < tNext) {// evolve from t to t_next
46              gsl_odeiv_evolve_apply(evolve, control, step,
47                  &system, &t, tNext, &tau0, y);
```

```
48        }
49            // save solution at t = t_next
50            file << scientific << setprecision(5) << setw(12)
51                << t << " " << setw(12) << y[0] << " "
52                << setw(12) << y[1] << endl;
53        }
54        // all done; free up the gsl_odeiv stuff
55        gsl_odeiv_evolve_free(evolve);
56        gsl_odeiv_control_free(control);
57        gsl_odeiv_step_free(step);
58        return 0;
59  }
```

We prescribe the initial time step; the program stops when the required relative or absolute accuracy is achieved.

7.2.5 Specification of the right-hand side

Here we also present the part of the program corresponding to the specification of the right-hand side for the system of equations (7.2), (7.3).

Listing 7.3.

```
1  // array of right-hand-side functions y[i] to be integrated.
2  int rhs(double, const double y[], double f[], void *params) {
3      double *al = (double *) params;
4      // right-hand-side functions at t
5      f[0] = al[0] * y[0] - al[1] * y[0] * y[1];
6      f[1] = - al[2] * y[1] + al[3] * y[0] * y[1];
7      return GSL_SUCCESS;
8  }
```

Some predictions for the predator-prey system visualized using gnuplot are given in the following.

7.2.6 Visualization script

The numerical results are written in the file lotka.dat for further visualization via gnuplot. To use this visualization system, we construct the script written in the file lotka.gnu; it is shown as follows:

```
#!/bin/gnuplot
set terminal png
set output "lotka.png"
set xlabel "t"
plot "lotka.dat" using 1:2 title "v" with lines, \
"lotka.dat" using 1:3 title "w" with lines
pause -1
```

The plot obtained after data processing is saved in the file `lotka.png`.

7.2.7 The basic variant of input data

Assume that the coefficients of the system of equations (7.2), (7.3) are the following:

$$\alpha_1 = 0.05, \quad \alpha_2 = 0.005, \quad \alpha_3 = 0.01, \quad \alpha_4 = 0.0002.$$

In this case, the stationary state is

$$\bar{w} = 10, \quad \bar{v} = 50.$$

The numerical solution of the problem (7.2), (7.3) with the initial conditions $v^0 = 10$ and $w^0 = 10$ is shown in Figure 7.3.

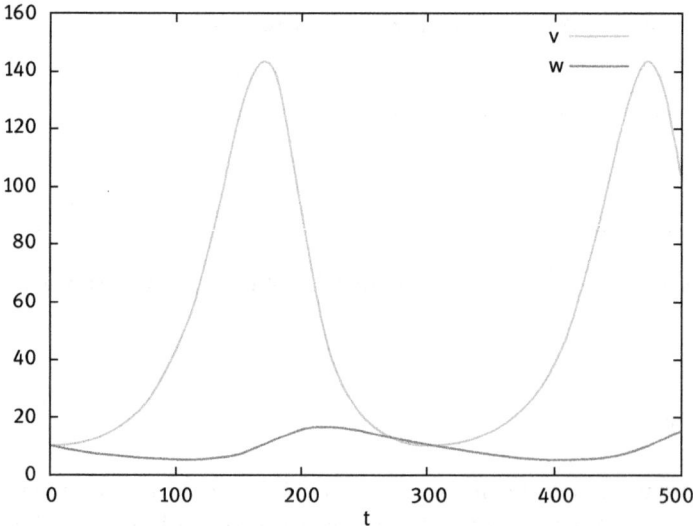

Fig. 7.3. Population dynamics for the predator-prey system.

7.2.8 Variation of parameters

Using the program presented above, we can study the behavior of the system varying parameters of the problem. In particular, it is interesting to analyze the influence of the initial conditions v^0, w^0 on the dynamics of the population.

For example, Figure 7.4 demonstrates the time-evolution of the number of species for the initial conditions $v^0 = 100$ and $w^0 = 10$.

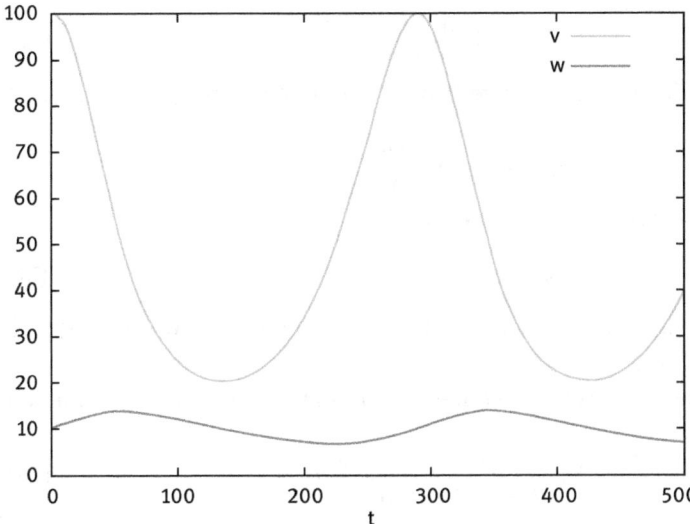

Fig. 7.4. Population dynamics for $v^0 = 100$ and $w^0 = 10$.

To continue this parametric study, we can also vary the coefficients α_1, α_2, α_3, α_3.

7.3 Water infiltration

In this section we consider a mathematical model describing water infiltration from the Earth's surface into the soil. It is based on a one-dimensional quasi-linear parabolic equation. The computational algorithm for studying this problem is presented below along with the numerical results.

7.3.1 The equation of water transport

The water precipitation falls on the Earth's surface and seeps into the soil within a certain depth. Denote by w the water content (soil moisture) and let w^* be the maximum water content ($0 \leq w \leq w^*$). To investigate the **water infiltration** in an unsaturated zone, we apply the one-dimensional equation of water transport. Assume that the axis z is directed into the ground, then the equation of water transport may be written as

$$\frac{\partial w}{\partial t} = \frac{\partial}{\partial z}\left(D(w)\frac{\partial w}{\partial z}\right) - \frac{\partial}{\partial z}k_w(w). \tag{7.5}$$

The coefficient of hydraulic conductivity k_w depends essentially on the water content, i.e. we have $k_w = k_w(w)$. For example, the following relation (S. F. Averyanov's formula) is typical:

$$k_w = k\left(\frac{w}{w^*}\right)^\sigma, \tag{7.6}$$

where k is the flow coefficient and σ ranges from 2 to 4 or more. In equation (7.5), we have

$$D(w) = -k_w(w)\frac{d\psi}{dw},$$

where ψ is the suction height. The dependence

$$\psi(w) = \alpha - \beta w \tag{7.7}$$

is widely used in predictions of problems like this. The coefficients α and β in (7.7) are determined experimentally.

Let $u = w/w^*$ be the relative water content; then, in view of (7.6), (7.7), equation (7.5) takes the form

$$\frac{\partial u}{\partial t} = k\beta\frac{\partial}{\partial z}\left(u^\sigma\frac{\partial u}{\partial z}\right) - \frac{k}{w^*}\frac{\partial}{\partial z}u^\sigma. \tag{7.8}$$

7.3.2 Self-similar solution

For the water infiltration process, the **self-similar solution** is the exact solution of equation (7.8) corresponding at the asymptotic stage to the motion of the wetting boundary with some constant speed v, i.e. it is a traveling wave-like self-similar solution.

Therefore, the function $\zeta = z - vt$ is selected as the self-similar variable. Some self-similar solutions of this type can be obtained in the explicit form. For instance, for $\sigma = 2, 3, 5$, we have

$$\zeta = \beta w^*(u + \ln(1 - u)) + \text{const}, \quad \sigma = 2,$$

$$\zeta = \beta w^*\left(u + \frac{1}{2}\ln((1 - u)/(1 + u)) + \text{const}, \quad \sigma = 3,\right.$$

$$\zeta = \beta w^*\left(u + \frac{1}{4}\ln((1 - u)/(1 + u)) - \frac{1}{2}\arctan u\right) + \text{const}, \quad \sigma = 5,$$

with $v = k/w^*$. This set of solutions allows us to study the influence of the parameter σ on a steady water profile.

Visualization of the self-similar solutions presented above is performed for const = 0. The corresponding script (file self.gnu) is as follows:

```
#!/usr/bin/gnuplot
set terminal png
set output "self.png"
t1 = 5./3
t2 = 10./3
t3 = 5
f1(x,y)=y+log(1-y) - x
f2(x,y)=y+log((1-y)/(1+y))/2 - x
f3(x,y)=y+log((1-y)/(1+y))/4 - atan(y)/2 - x
set contour
set isosamples 1000,1000
unset key
set cntrparam level 0
unset surface
unset clabel
set view map
set size ratio 1/2
set xrange [0:5]
set yrange [0:1]
set xlabel "z"
set ylabel "u"
splot f1(x,y) + t1 w lines lc rgb "blue", \
f1(x,y) + t2 w lines lc rgb "blue", \
f1(x,y) + t3  w lines lc rgb "blue", \
f2(x,y) + t1 w lines lc rgb "green", \
f2(x,y) + t2 w lines lc rgb "green", \
f2(x,y) + t3 w lines lc rgb "green", \
f3(x,y) + t1 w lines lc rgb "red", \
f3(x,y) + t2 w lines lc rgb "red", \
f3(x,y) + t3 w lines lc rgb "red
pause -1
```

The self-similar solutions for $\sigma = 2, 3, 5$ and $\beta w^*, v = 1$ are depicted at time moments $t = 5/3, t = 10/3$ and $t = 5$ in Figure 7.5.

Now we shall discuss the computational algorithm for studying this problem and present some numerical results.

7.3.3 Difference scheme

To solve numerically the boundary value problem for water transport, we apply the finite-difference scheme with the coefficients taken from the previous time level. The numerical implementation is based on the solution of a system of linear equations with a tridiagonal matrix.

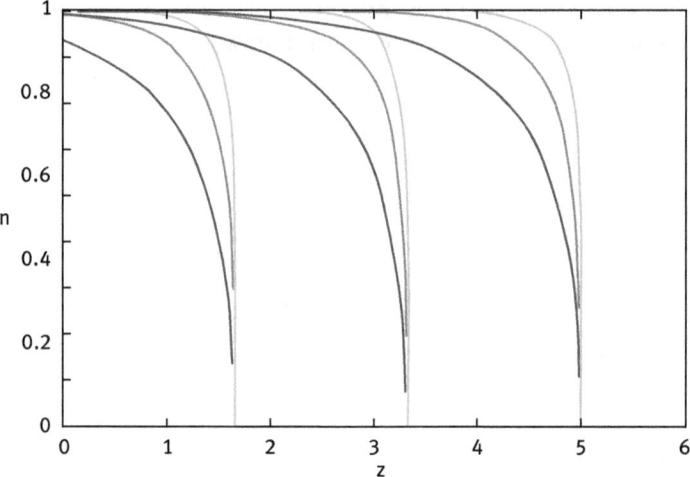

Fig. 7.5. Water profile dynamics for $\sigma = 2$ (black line), $\sigma = 3$ (gray) and $\sigma = 5$ (silver).

We seek the solution of the water infiltration problem where the soil is initially unsaturated. We restrict ourselves to the case with $k\beta = 1$, $k/w^* = 1$. The equation (7.8) is considered for $0 < z < l$ with the following boundary and initial conditions:

$$u(0, t) = 1, \quad u(l, t) = 0, \quad 0 < t < T,$$

$$u(z, 0) = 0, \quad 0 < z < l.$$

For the numerical solution of the boundary value problem, we introduce a **uniform mesh** on the segment $[0, l]$:

$$\bar{\omega} \equiv \omega \cup \partial\omega = \{z \mid z = z_i = ih, \quad i = 0, 1, \dots, N, \quad Nh = l\},$$

where ω is the set of interior nodes ($i = 1, 2, \dots, N - 1$). Denote the approximate solution at the time moment $t^n = n\tau$ by y^n, where τ is the time step. We employ the linearized difference scheme with coefficients taken from the previous time level. For the interior nodes of the spatial grid and the time moments $n = 0, 1, \dots$, the **difference scheme** has the form

$$\frac{y_i^{n+1} - y_i^n}{\tau} - \frac{1}{h^2}\left(\frac{y_{i+1}^n + y_i^n}{2}\right)^\sigma (y_{i+1}^{n+1} - y_i^{n+1}) + \frac{1}{h^2}\left(\frac{y_i^n + y_{i-1}^n}{2}\right)^\sigma (y_i^{n+1} - y_{i-1}^{n+1})$$

$$+ \sigma(y_i^n)^{\sigma-1}\frac{y_{i+1}^{n+1} - y_{i-1}^{n+1}}{2h} = 0, \quad i = 1, 2, \dots, N - 1.$$

The boundary conditions appear as

$$y_0^{n+1} = 1, \quad y_N^{n+1} = 0.$$

7.3.4 Program

In the following we present the code for solving the boundary value problem for the equation of water transport. A fine enough mesh in space ($N = 100$) is used in calculations conducted with the coefficient $\sigma = 2$.

Listing 7.4.

```
1   // file: bvpt.cpp
2   #include <iostream>
3   #include <iomanip>
4   #include <fstream>
5   #include <math.h>
6   #include <gsl/gsl_matrix.h>
7   #include <gsl/gsl_vector.h>
8   #include <gsl/gsl_linalg.h>
9   using namespace std;
10  int main() {
11      const int sigma = 2;
12      const double l = 6.0;
13      const double T = 5.0;
14      const int N = 100;
15      double h = l / N;
16      double tau = 0.01;
17      int nT = 0;
18      int nWr = 50;
19      gsl_vector * u = gsl_vector_alloc(N + 1); // solution
20      gsl_vector * a = gsl_vector_alloc(N); // sub-diagonal
21      gsl_vector * b = gsl_vector_alloc(N + 1); // diagonal
22      gsl_vector * c = gsl_vector_alloc(N); // super-diagonal
23      gsl_vector * f = gsl_vector_alloc(N + 1); // rhs
24      // initial data
25      int i;
26      double t = 0.0;
27      for (i = 1; i < N + 1; i++) {
28          gsl_vector_set(u, i, 0.0);
29      }
30      gsl_vector_set(u, 0, 1.0);
31      ofstream file("bvpt.dat");
32      //  time cycle
33      double y, yl, yr;
34      double kl, kr, cv;
35      while (t < T - 0.5 * tau) {
36          t += tau;
37          // internal nodes
38          for (i = 1; i < N; i++) {
39              y = gsl_vector_get(u, i);
40              yl = gsl_vector_get(u, i - 1);
```

```
41          yr = gsl_vector_get(u, i + 1);
42          kl = pow((yl + y) / 2, sigma) * tau / (h * h);
43          kr = pow((yr + y) / 2, sigma) * tau / (h * h);
44          cv = sigma * pow(y, sigma - 1) * tau / (2 * h);
45          gsl_vector_set(a, i - 1, -kl - cv);
46          gsl_vector_set(b, i, kl + kr + 1.0);
47          gsl_vector_set(c, i, -kr + cv);
48          gsl_vector_set(f, i, y);
49      }
50      // boundary nodes
51      gsl_vector_set(b, 0, 1.0);
52      gsl_vector_set(c, 0, 0.0);
53      gsl_vector_set(f, 0, 1.0);
54      gsl_vector_set(a, N - 1, 0.0);
55      gsl_vector_set(b, N, 1.0);
56      gsl_vector_set(f, N, 0.0);
57      gsl_linalg_solve_tridiag(b, c, a, f, u);
58      nT++;
59      if (nT == nWr) { // save solution
60          file << "#t = " << setw(12) << t << endl;
61          for (i = 0; i < N; i++) {
62              file << scientific << setprecision(5)
63                  << setw(12) << i * h
64                  << setw(12) << gsl_vector_get(u,
65                  << endl;
66          }
67          nT = 0;
68          file << endl;
69      }
70  }
71  gsl_vector_free(u);
72  gsl_vector_free(a);
73  gsl_vector_free(b);
74  gsl_vector_free(c);
75  gsl_vector_free(f);
76  return 0;
77 }
```

As above, the visualization of the predictions is performed by `gnuplot`.

7.3.5 Visualization

The results at the specified time moments (at the interval of `nWr` = 50 steps) are written in the file `bvpt.dat`. The script for `gnuplot` is contained in the file `bvpt.gnu` and has the following form:

```
#!/usr/bin/gnuplot
plot "bvpt.dat" using 1:2 title "u"  with lines
pause -1
```

The plots of the solution at the individual time moments (multiples of $\delta t = 0.5$) are written in file bvpt.png.

7.3.6 Predictions

The dynamics of the infiltration process for the basic case is shown in Figure 7.6. We can observe that during the time-evolution process the infiltration front transforms practically to the self-similar solution (see Figure 7.5 for a comparison).

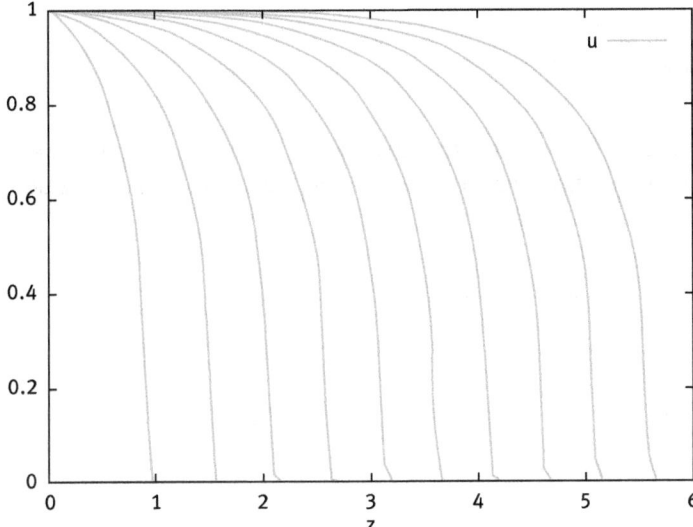

Fig. 7.6. Water dynamics with the constant soaking on the Earth's surface ($u(0, t) = 1$).

A more complicated situation, when soaking on the Earth's surface is varying in time, is reflected in Figure 7.7. In this case, until the time moment $0.5T$, the soaking exists and is constant, and after this point the soaking is absent. This situation is modeled by specifying the boundary condition at $z = 0$ in the form

$$u(0, t) = \begin{cases} 1, & 0 < t < 0.5T, \\ 0, & 0.5T < t < T. \end{cases}$$

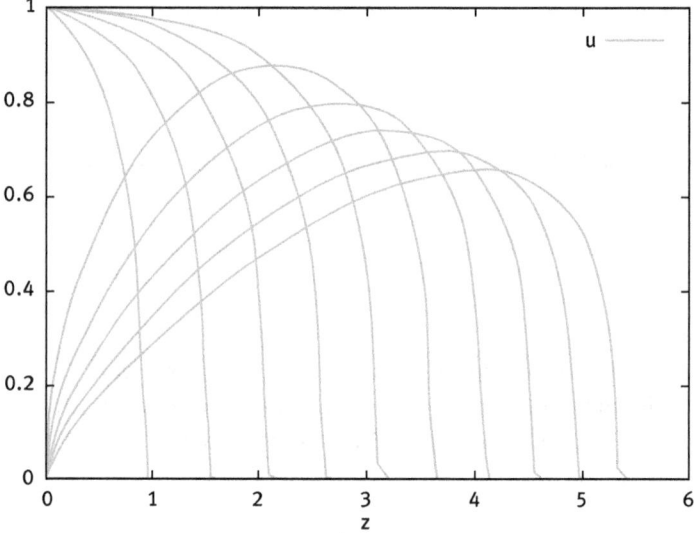

Fig. 7.7. The case with soaking varying in time.

The influence of the parameter σ is demonstrated in Figure 7.8. As for the self-similar solutions, the increase of σ results in sharping the soaking front (increasing its gradient).

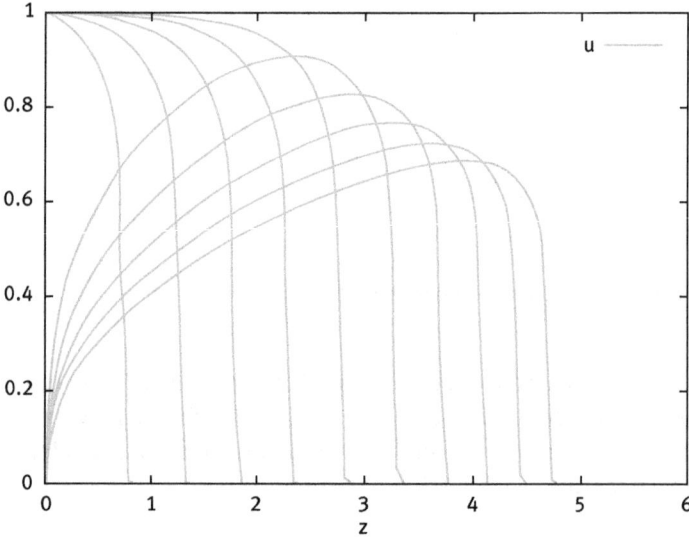

Fig. 7.8. Water dynamics for $\sigma = 3$.

7.4 Torsion of cylindrical bars

The torsion of cylindrical bars is the classical problem of the theory of elasticity. In this section, we present a computational algorithm for solving this problem and discuss the results obtained.

7.4.1 Problem formulation

The problems of the **elastic torsion of cylindrical bars** belong to the class of the basic and well-studied problems of solid mechanics. To study the torsion of bars with simply connected cross-sections of an arbitrary shape, it is necessary to solve the Dirichlet problem for the two-dimensional Poisson equation. These boundary value problems present the class of the basic problems of mathematical physics, which are provided with the well-known numerical methods.

We consider a cylindrical bar with a cross-section D in a plane x_1, x_2 (the axis x_3 is directed along the axis of the bar) subjected to uniform torsion. Let α be the swirl angle per unit length of the bar. For the components of the stress tensor, we have

$$\tau_{11} = \tau_{22} = \tau_{33} = \tau_{12} = 0, \quad \tau_{13} = \frac{\partial u}{\partial x_2}, \quad \tau_{23} = -\frac{\partial u}{\partial x_1}, \tag{7.9}$$

where $u = u(x)$, $x = (x_1, x_2)$. For elastic deformation, the stress function $u(x)$ satisfies the equation

$$\frac{\partial \tau_{13}}{\partial x_2} - \frac{\partial \tau_{23}}{\partial x_1} = -2\alpha G, \quad x \in D, \tag{7.10}$$

where G is the shear modulus of bar material. From (7.9), (7.10), we arrive at the Poisson equation

$$-\frac{\partial^2 u}{\partial x_1^2} - \frac{\partial^2 u}{\partial x_2^2} = 2\alpha G, \quad x \in D. \tag{7.11}$$

Assume that the section D of the bar is simply connected. Then equation (7.11) is complemented by the homogeneous Dirichlet condition on the boundary:

$$u(x) = 0, \quad x \in \partial D. \tag{7.12}$$

In processing the computational results, the focus is on the torque

$$M = 2 \int_D u(x) dx.$$

Multiplying equation (7.9) by $u(x)$ and integrating the resulting equation over the domain Ω, we get

$$M = \frac{1}{\alpha G} \int_D w(x) dx,$$

where the function

$$w(x) = \left(\frac{\partial u}{\partial x_1}\right)^2 + \left(\frac{\partial u}{\partial x_1}\right)^2$$

defines local features of the loads.

7.4.2 Difference problem

To solve numerically the Dirichlet problem (7.11), (7.12), we will consider the problem in a rectangle Ω completely containing the domain D ($D \subset \Omega$). Thus, we have

$$\Omega = \{x \mid x = (x_1, x_2),\ 0 < x_\alpha < l_\alpha,\ \alpha = 1, 2\}.$$

In Ω, we introduce a uniform mesh $\bar{\omega} = \omega \cap \partial\omega$ with steps h_1 and h_2 in the corresponding direction, respectively. Let ω be the set of interior nodes, i.e.

$$\omega = \{x \mid x = (x_1, x_2),\ x_\alpha = i_\beta h_\alpha,\ i_\beta = 1, 2, \ldots, N_\alpha,\ N_\alpha h_\alpha = l_\alpha,\ \alpha = 1, 2\},$$

and $\partial\omega$ is the set of boundary nodes. The difference solution of the problem (7.11), (7.12) is denoted by $y(x), x \in \bar{\omega}$.

For the nodes outside the computational domain, in view of the boundary condition (7.12), we assume

$$y(x) = 0, \quad x \in \bar{\omega}, \quad x \notin D. \tag{7.13}$$

At other nodes of the spatial grid, equation (7.11) is approximated by the following relations:

$$-\frac{1}{h_1}\left(\frac{y(x_1 + h_1, x_2) - y(x)}{h_1} - \frac{y(x) - y(x_1 - h_1, x_2)}{h_1}\right)$$

$$-\frac{1}{h_2}\left(\frac{y(x_1, x_2 + h_2) - y(x)}{h_2} - \frac{y(x) - y(x_1, x_2 - h_2)}{h_2}\right) = 2\alpha G, \quad x \in \omega, \quad x \in D. \tag{7.14}$$

The difference scheme (7.13), (7.14) results from the piecewise linear approximation of the boundary of the computational domain D with grid nodes located on the boundary.

7.4.3 Numerical algorithm and program

To find the difference solution of the problem, the **iterative successive over-relaxation (SOR) method** is used.

Let $y^n(x)$ be the approximate solution at the n-th iteration and $0 < \sigma < 2$ be a relaxation parameter; then

$$\left(\frac{2}{h_1^2} + \frac{2}{h_2^2}\right) y^{n+1}(x) = (1 - \sigma) \left(\frac{2}{h_1^2} + \frac{2}{h_2^2}\right) y^n(x)$$

$$+ \sigma \left(\frac{1}{h_1^2} y^{n+1}(x_1 - h_1, x_2) + \frac{1}{h_2} y^{n+1}(x_1, x_2 - h_2)\right.$$

$$\left. + \frac{1}{h_1^2} y^n(x_1 + h_1, x_2) + \frac{1}{h_2} y^n(x_1, x_2 + h_2) + 2\alpha G\right), \quad x \in \omega, \quad x \in D.$$

The program for solving the torsion problem is as follows:

Listing 7.5.

```cpp
// file: tors.cpp
#include <iostream>
#include <iomanip>
#include <fstream>
#include <math.h>
using namespace std;
#define n1 200
#define n2 100
int main() {
    double tol, r, er = 0.0000001;
    double omega = 1.9; // The SOR parameter
    double u[n1 + 1][n2 + 1];
    int i, j, it, itMax = 1000;
    double l1 = 2.0;
    double l2 = 1.0;
    double g = 2.0;
    double h1 = l1 / n1;
    double h2 = l2 / n2;
    // initial
    it = 0;
    for (i = 0; i < n1 + 1; i++)
        for (j = 0; j < n2 + 1; j++)
            u[i][j] = 0.0;
    // iterations circle
    double a1 = 1.0 / (h1 * h1);
    double a2 = 1.0 / (h2 * h2);
    tol = 1.0;
    while (tol > er && it <= itMax) {
        tol = 0.0;
        for (i = 1; i < n1; i++) {
            for (j = 1; j < n2; j++) {
                if (i * h1 < l1 / 2 && j * h2 < l2 / 2) continue;
                else {
                    r = ((u[i + 1][j] + u[i - 1][j]) * a1
```

```
35                          + (u[i][j + 1] + u[i][j - 1]) * a2 + g)
36                        / (2 * (a1 + a2)) - u[i][j];
37                    u[i][j] += omega * r;
38                    if (fabs(r) > tol)
39                        tol = fabs(r);
40                }
41            }
42        }
43        it++;
44    }
45    printf("Number of iteration = %u, tolerance = %g\n", it, tol);
46    // save solution
47    ofstream file("tors.dat");
48    for (i = 0; i < n1 + 1; i++) {
49        for (j = 0; j < n2 + 1; j++)
50            file << scientific << setprecision(5) << setw(12)
51                << i*h1            << " " << setw(12) << j*h2
52                << " " << setw(12) << u[i][j] << endl;
53        file << endl;
54    }
55    double v;
56    ofstream file1("tors1.dat");
57    for (i = 1; i < n1 ; i++) {
58        for (j = 1; j < n2 ; j++) {
59            v = (u[i+1][j] - u[i-1][j])*(u[i+1][j] - u[i-1][j])*a1
60                + (u[i][j+1] - u[i][j-1])*(u[i][j+1] - u[i][j-1])*a2;
61            file1 << scientific << setprecision(5) << setw(12)
62                << i*h1 << " " << setw(12) << j*h2
63                << " " << setw(12) << sqrt(v) << endl;
64        }
65        file1 << endl;
66    }
67    return 0;
68 }
```

In the particular case presented below, the torsion problem is studied for the bar with the rotated L-shape cross-section, i.e. the cross-section is the rectangle ($l_1 = 2$, $l_2 = 1$) without the rectangular part

$$\Omega \setminus D = \left\{ x \mid x = (x_1, x_2), \ 0 < x_\alpha < \frac{1}{2}l_\alpha, \ \alpha = 1, 2 \right\}.$$

The main calculation parameters are the following. The uniform mesh with 201×101 nodes is used. The relaxation parameter σ is equal to 1.9. The iterative process stops as soon as the maximum residual becomes less then 1×10^{-7}.

7.4.4 Numerical results

The numerical solution is written to the file `tors.dat`. In addition, the function $w(x)$ is also calculated at the interior nodes of the mesh (the file `tors1.dat`). For visualization, the script (the file `tors.gnu`) is employed.

```
#!/usr/bin/gnuplot
set terminal png
set output "tors.png"
set pm3d map
set palette rgb 23,28,3 negative
splot "tors.dat"
set output "tors1.png"
splot "tors1.dat"
pause -1
```

During the run of the program, the number of iterations and tolerance are printed:

```
Number of iteration = 518, tolerance = 9.94416e-08
```

The predicted solution is shown in Figure 7.9 as colored values of the stress function $u(x)$ with the legend at the right. Figure 7.10 presents the calculated function $w(x)$ in the same style.

Using this program, it is possible to investigate the influence of the mesh size N_1, N_2 on the accuracy of the numerical solution as well as the impact of the relaxation parameter σ on the convergence rate. We might also consider the torsion problem for bars with other cross-sections.

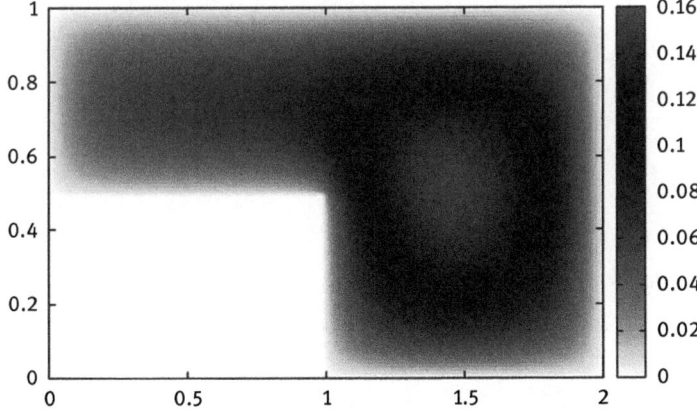

Fig. 7.9. Values of the stress function $u(x)$.

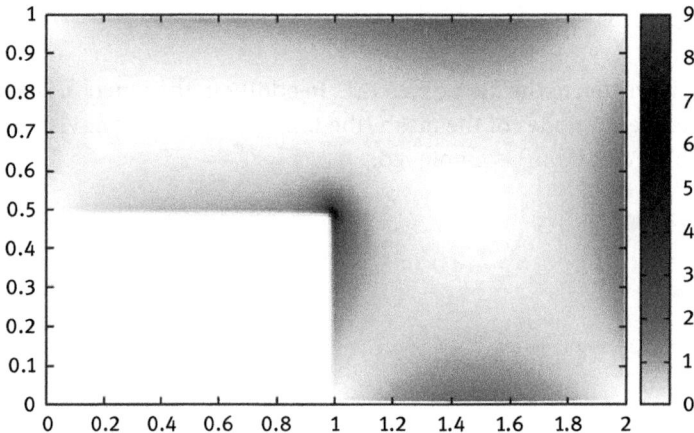

Fig. 7.10. Values of the function $w(x)$.

Bibliography

[1]   D. Abbott, *Embedded Linux Development Using Eclipse*, Newnes, Burlington, MA, 2009.

[2]   D. Boffi, F. Brezzi and M. Fortin, *Mixed Finite Element Methods and Applications*, Springer-Verlag, Berlin, 2013.

[3]   E. Burnette, *Eclipse IDE Pocket Guide*, O'Reilly Media, Sebastopol, CA, 2005.

[4]   D. Carlson, *Eclipse Distilled*, Addison-Wesley, Upper Saddle River, NJ, 2005.

[5]   G. Connan and S. Grognet, *Guide du Calcul avec les logiciels libres*, Dunon, Paris, 2008 (French).

[6]   A. Ern and J.-L. Guermond, *Theory and Practice of Finite Elements*, Springer, New York, NY, 2004.

[7]   M. Galassi, J. Davies, J. Theiler, B. Gough, G. Jungman, P. Alken, M. Booth, F. Rossi and R. Ulerich, *GNU Scientific Library: Reference Manual Version 1.16*, The GSL Team, www.gnu.org/software/gsl/, 2013.

[8]   G. N. Gatica, *A Simple Introduction to the Mixed Finite Element Method: Theory and Applications*, Springer, Cham, 2014.

[9]   B. Gough, *An Introduction to GCC: for the GNU Compilers gcc and g++*, Network Theory Ltd., Bristol, UK, 2004.

[10] A. Griffith, *GCC: The Complete Reference*, McGraw-Hill/Osborne, New York, NY, 2002.

[11] C. Grossmann, H.-G. Roos and M. Stynes, *Numerical Treatment of Partial Differential Equations*, Springer-Verlag, Berlin, 2007.

[12] R. Heathfield, L. Kirby, I. Woods, S. Summit, I. Kelly and et al., *C Unleashed*, Sams, Indianapolis, IN, 2000.

[13] P. K. Janert, *gnuplot in Action: Unerstanding Data with Graphs*, Manning Publications Co., Greenwick, UK, 2010.

[14] J. Kampf, *Ocean Modelling for Beginners: Using Open-Source Software*, Springer-Verlag, Berlin, 2009.

[15] J. Kampf, *Advanced Ocean Modelling: Using Open-Source Software*, Springer-Verlag, Berlin, 2010.

[16] B. W. Kernigan and D. M. Ritchie, *The C Programming Language*, Prentice Hall, Englewood Cliffs, NJ, 1988.

[17] B. Klemens, *21st Century C*, O'Reilly Media, Sebastopol, CA, 2013.

[18] S. Koranne, *Handbook of Open Source Tools*, Springer, New York, NY, 2011.

[19] M. G. Larson and F. Bengzon, *The Finite Element Method: Theory, Implementation, and Applications*, Springer-Verlag, Berlin, 2013.

[20] S. Lui, *Numerical Analysis Of Partial Differential Equations*, John Wiley & Sons, Hoboken, NJ, 2011.

[21] P. Moin, *Fundamentals of Engineering Numerical Analysis*, Cambridge University Press, New York, NY, 2010.

[22] L. Phillips, *gnuplot Cookbook*, Packt Publishing Ltd., Birmingham, UK, 2012.

[23] J. Pitt-Francis and J. Whiteley, *Guide to Scientific Computing in C++*, Computing in C++, London, 2012.

[24] S. Prata, *C Primer Plus*, Sams, Indianapolis, IN, 2004.

[25] E. Scheinerman, *C++ for Mathematicians: An Introduction for Students and Professionals*, Chapman & Hall/CRC, Boca Raton, FL, 2006.

[26] H. Schildt, *C++ from the Ground Up*, McGraw-Hill/Osborne, New York, NY, 2003.

[27] H. Schildt, *C++: The Complete Reference*, McGraw-Hill/Osborne, New York, NY, 2003.

[28] H. Schildt, *Herb Schildt's C++ Programming Cookbook*, McGraw-Hill, New York, NY, 2008.

[29] D. Scuset, *Eclipse 3.1*, Dept. Computer Science, University of Manitoba, 2012.

[30] Y. Shapira, *Solving PDEs in C++: Numerical Methods in a Unified Object-Oriented Approach*, SIAM, Philadelphia, PA, 2006.

[31] Y. Shapira, *Mathematical objects in C++: Computational Tools in a Unified Object-Oriented Approach*, CRC Press, Boca Raton, FL, 2009.

[32] B. Stroustrup, *The C++ Programming Language*, Addison-Wesley, Reading, MA, 1997.

[33] B. Stroustrup, *Programming: Principles and Practice Using C++*, Addison-Wesley, Upper Saddle River, NJ, 2009.

[34] B. Stroustrup, *A Tour of C++*, Addison-Wesley, Upper Saddle River, NJ, 2014.

[35] B. Szabo and I. Babushka, *Introduction to Finite Element Analysis: Formulation, Verification and Validation*, John Wiley & Sons, Chichester, UK, 2011.

[36] J. A. Trangenstein, *Numerical Solution of Elliptic and Parabolic Partial Differential Equations*, Cambridge University Press, New York, NY, 2013.

[37] K. Velten, *Mathematical Modeling and Simulation: Introduction for Scientists and Engineers*, Wiley-VCH Verlag GmbH & Co. KGaA, Weinheim, FRG, 2009.

[38] W. von Hagen, *The Definitive Guide to GCC*, Apress, Berkeley, CA, 2006.

Index

Array, 13
– multidimensional, 15
– two-dimensional, 15

Basic Linear Algebra Subprograms (BLAS), 120
Boolean variable, 31
Breakpoints, 72

Class, 32
– constructor, 35
– definition, 32
– derived, 38
– destructor, 35
– method, 33
– object, 34
– string, 49
Compiler, 60
Cygwin, 52

Data type, 3
Difference scheme, 222

Eclipse IDE, 81
– Editor, 93
– Building a project, 93
– Creating a project, 88
– debugger, 99
– installation, 82
– perspective, 86
– plug-ins, 83
Elastic torsion of cylindrical bar, 227

File, 22
Format string, 23
Function, 17
– atof, 48
– atoi, 48
– calloc, 31
– clock, 48
– close, 46
– ctime, 47
– eof, 46
– free, 31
– friend, 37
– malloc, 31

– open, 46
– overloaded, 36
– scanf, 24
– time, 47
– virtual, 40

GNU Compiler Collection (GCC), 51
GNU Scientific Library (GSL), 103
Gnuplot
– axis, 203
– data blocks, 189
– graph decoration, 202
– grid lines, 207
– installation, 182
– labels, 206
– legend, 202
– line style, 201
– plot command, 183
– script, 188
– smooth option, 191
– styles, 197
– type of the terminal, 195
GSL
– B-splines, 170
– basic statistical functions, 108
– block, 118
– Cauchy problem for system of ODEs, 139
– Chebyshev approximation, 176
– combination, 112
– eigenvalues and eigenvectors, 127
– evolution function, 143
– explicit, implicit, and multistep methods, 140
– fast Fourier transform, 174
– install, 103
– interpolation, 164
– least squares method, 167
– LU decomposition, 125
– matrix, 119
– minimization of multidimensional functions, 160
– minimization of one-dimensional functions, 157
– Monte Carlo integration methods, 136
– multidimensional root-finding, 154
– multiset, 117

– numerical derivative, 129
– numerical integration, 132
– permutation, 110
– physical constants, 105
– polynomials, 149
– QR decomposition, 127
– random number generators, 106
– roots of one-dimensional function, 151
– standard mathematical functions, 104
– vector, 118

IDE
– Code::Blocks, 80
– CodeLite, 81
– Eclipse, 81
– Eclipse CDT C/C++ Development Tools, 81
– Geany, 80
– NetBeans, 80
– Qt Creator, 80
Input files
– C, 65
– C++, 65
Integrated development environment (IDE), 79
Iterative successive over-relaxation (SOR)
        method, 228

Least squares method, 209
Library
– ctime, 47
– ifstream, 46
– iostream, 41
– ofstream, 46
– shared (dynamic), 68
– static, 68
Linux
– info pages, 56
– man (manual) pages, 55
– quick reference, 54
Lotka-Volterra equations, 213

Makefile, 74
– phony target, 77
– sections, 76
– variables, 77
Memory allocation, 31

Minimalist GNU for Windows (MinGW), 53

One-step Runge–Kutta method, 215
Operator, 6
– arithmetic, 6
– logical, 7
– overloaded, 37
– relational, 7
Options
– control a language, 66
– optimization, 67
– warnings, 66

Pointer, 15
Polynomial approximation, 209
Preprocessor directive, 25
Program
– g++, 62
– gcc, 60
– GNU Debugger (GDB), 70
– gnuplot, 181
– Konqueror, 58
– make, 74
– yelp, 58
Programming language, 1
– C, 1
– C++, 29
Protected element, 39

Runge–Kutta–Fehlberg method, 214

Self-similar solution, 220
Standard library of C++, 47
Structure, 20

Type conversion, 8

Ubuntu, 52
– Advanced packaging tool (apt), 52
– Synaptic Package Manager, 52
– Ubuntu Software Center, 52
Uniform mesh, 222

Water infiltration, 220
Windows, 52